Energy Management in Buildings

The role of the energy manager has evolved significantly as the task of cutting greenhouse gas emissions from buildings has become increasingly important. Managers are now technical experts, negotiators, construction project managers, procurement specialists and efficiency advocates, and often provide energy services to others.

This comprehensive book covers how to:

- conduct an energy audit
- plan a monitoring and verification strategy
- make any energy-saving campaign successful
- evaluate and make the financial case for energy-saving measures
- make use of free energy for lighting and managing heat loss and gain.

It also contains special chapters on:

- ventilation, heating and cooling
- demand management through automated systems
- lighting
- most requirements of industrial facilities
- the management of data centres
- regulatory requirements in Britain, Europe and the United States
- the use of smart meters and monitoring
- how to achieve zero energy buildings
- the use of renewable energy.

For all professional energy, building and facilities managers, energy consultants, students, trainees and academics, this book takes the reader from basic concepts to the latest advanced thinking, with principles applicable anywhere in the world and in any climate.

David Thorpe is community manager of Sustainable Cities Collective, an online community for leaders of major metropolitan areas, urban planning and sustainability professionals. Until 2013 he was for 13 years News Editor of *Energy and Environment Management* magazine, for which website he also wrote a weekly op-ed column. He is also the author of several books and hundreds of articles on related subjects. Formerly director of publications at the Centre for Alternative Technology, he has written two other books in the Earthscan Expert series, *Sustainable Home Refurbishment* and *Solar Technology*, and several B2B ebooks for the publisher DoSustainability. He runs his own sustainable media consultancy, Cyberium, manages the Green Deal Advice website, and blogs regularly as The Low Carbon Kid. Find him on Twitter @DavidKThorpe.

Earthscan Expert Series
Series editor: Frank Jackson

Energy Management in Buildings

The Earthscan Expert Guide

David Thorpe

Series Editor:
Frank Jackson

Routledge
Taylor & Francis Group

LONDON AND NEW YORK

earthscan
from Routledge

First published 2014
by Routledge
2 Park Square, Milton Park, Abingdon, Oxon OX14 4RN

and by Routledge
711 Third Avenue, New York, NY 10017

Routledge is an imprint of the Taylor & Francis Group, an informa business

British Library Cataloguing in Publication Data
A catalogue record for this book is available from the British Library

Library of Congress Cataloging in Publication Data has been requested.

ISBN: 978-0-415-70646-9 (hbk)
ISBN: 978-1-315-88475-2 (ebk)

Typeset in Sabon by Keystroke, Station Road, Codsall, Wolverhampton

Printed in Great Britain by Bell & Bain Ltd, Glasgow

Contents

Illustrations

Figures

Tables

Preface

It's amazing how long it takes for some things to catch on, even if they're great ideas. I have been the News Editor of *Energy and Environment Management* magazine since 2000. Before that, I worked at the world-renowned Centre for Alternative Technology in Wales, and even earlier for organisations such as Greenpeace, who commissioned a book about climate change not long after it was 'discovered'. That's about a quarter of a century ago.

This, and its companion volume *Energy Management in Industry*, is therefore a distillation of everything I have learned during this time. An astonishing amount of it is not new. I have seen so many attempts to try to get organisations and businesses to save energy. The arguments are always compelling: profits become increased as costs reduce. In the 1990s, the UK government's Energy Efficiency Best Practice Programme produced many detailed sector-specific how-to fact sheets. This continued when it became Action Energy, and then the Carbon Trust, which still exists, and to whom I am grateful for some of the illustrations provided in this volume. Many companies and organisations have implemented the recommendations of these, and other bodies, and discovered huge benefits that are not just to do with saving money.

But the mystery is why they have remained for so long in a small minority. So much more is possible. Renewable energy receives lots of headlines, but the most cost-effective carbon savings are achieved through energy efficiency in organisations.

There are many different types of organisations and businesses, public and private, and many different types of buildings. This book attempts to address all of these, from small businesses to office blocks, leased properties (including accommodation), hospitals, educational and service buildings. It argues that serious energy, building and facility managers should become accredited with the international standard for energy management, ISO 50001. There are plenty of training courses leading to this qualification. This book is intended to complement them, the standard and similar standards (see Introduction, Table 0.1).

Yes, it's questionable whether the kinds of employee motivational campaigns discussed in Chapter 3 are able to achieve lasting successes, but they are based on actual examples. I tend to believe that changes which take choice out of people's hands, so that efficiency is 'hard-wired' into a process, are more likely to produce reliable results. But behaviour change is still important, particularly in offices, educational institutions and hospitals. It is often said that we already have the technology to slash carbon emissions and live sustainably; but it's the psychological and behavioural aspect of this massive cultural change which is the real challenge.

Other work which I have done has hammered home the importance of inspiring board-level enthusiasm for sustainability. Without this, efforts will fail. I come across report after report – for example, from the Carbon Disclosure Project and from the World Business Council for Sustainable Development – saying that organisations which adopt low or zero carbon and sustainability targets as a core part of their business strategy perform above average in other respects. I believe the world needs to hear these messages.

It is a legitimate question whether it is good for the planet for business to save money on energy, since it will only reinvest it on using more resources. This book is not going to change capitalism, and instead argues for energy efficiency to be seen in the context of resource efficiency and, ultimately, a closed-loop system of production.

Change comes from two directions: people campaigning at the grassroots level, and from businesses and institutions. Both put pressure on politicians, who, unlike them, do not look 20 or 30 years ahead. Because investment cycles take 15 years at least, business is far more scared of climate change than politicians. (Although I write this from within the first country in the world to have sustainable development written into its constitution.) And business is well aware of resource scarcity. The institutions and businesses that will be around in 20 or 40 years' time will be the ones which have taken this message on board, who concentrate on getting more from less. It is good to get business to minimise its impact as much as possible and to be aware of the issues. It is business that is developing and investing in clean technology and it is clean technology that the world needs.

So I hope that these books, and the series of which they form a part, serve some use. I wish every energy manager, whether fledgeling newbies and students, or jaded veterans picking up new tricks, the best of luck. If there is anything they wish to include in a future edition, please would they get in touch.

I would like to thank all the energy managers who responded to requests for interviews to give the benefit of their personal experience to readers: Ashley Baxter, Lisa Gingell, Phil Bilyard, Andrew Bray, James White, Samantha Dean, Robert Kelk and Kit Oung. This invaluable advice may be found in between many of the chapters, and the interviews are included to show the variety of work an energy manager can be involved with, and to underline the point that it is people who are just as important, if not more so, as technology and data in securing energy savings, a fact which is often overlooked. I would also like to thank Alan Aldridge (Director of the Energy Services and Technology Association, ESTA), Dr John Ryan (Director of Certification Europe), as well as others who have commented on drafts of the book, including especially the Series Editor Frank Jackson; also Jo Abbess, Martin Kemp and Helen Adam, as well as the past editors of *Energy and Environment Management* magazine, including Nick Bent, who was Editor until recently.

David Thorpe, Wales, February 2013

Introduction

Energy managers: the hidden warriors

Energy managers are the secret warriors of the twenty-first century. Together they comprise a vital phalanx of the collective army whose essential role is to defend the world from destruction by climate change. For the most part unseen and unnoticed by the public, they toil in buildings everywhere, from hospitals to hotels, factories to data centres, from office blocks to leisure centres. After all, the energy used in buildings forms about 40 per cent of all energy used and 36 per cent of the world's CO_2 emissions.

Energy managers see things that are invisible to the majority. Their training leads them to sense the hidden flows of energy as it courses through pipes, wires, spaces and materials. They don't perceive a static situation, such as a boiler switched on, a light glowing, the window open, a tap dripping. They see this as part of a set of processes through time, visualising it as a series of transformations from one type of energy to another, such as, to take the example of a motor, from electricity to kinetic energy to dissipated heat energy.

For them, saving energy is eternal delight, in an evolution of the visionary poet William Blake's famous aphorism, 'energy is eternal delight'. Consequently these heroes are constantly struggling against the limits of the second law of thermodynamics, striving to prevent useful thermal or electrical energy from being dissipated irreversibly. Their catechism derives solely from the *primum movens* that 'No process is possible in which the sole result is the absorption of heat from a reservoir and its complete conversion into work'.

Energy efficiency

Energy efficiency is frequently described as the 'low hanging fruit'. The sector is expanding at a rate of 5 per cent per year. It is estimated that the global market value of innovative products in this sector could reach around £488 billion by 2050, and that, on average, most organisations can easily save at least 28 per cent of their energy costs with low-cost actions.

In the UK alone, innovative energy-saving measures in non-domestic buildings could save 18 MtCO$_2$ by 2020 and 86 MtCO$_2$ by 2050, depending upon the rate at which the measures can be deployed.[1] In the USA, the American Energy Manufacturing Technical Corrections Act was passed at the end of 2012, a modification of the Enabling Energy Savings Innovations Act. This promises to produce a boom in the sector. The US market for energy efficiency and services topped US$5.1 billion in 2011, according to Pike Research, and is now expected to reach US$16 billion in sales by 2020.

Figure 0.1 The context for energy efficiency within other concepts.

Source: Author

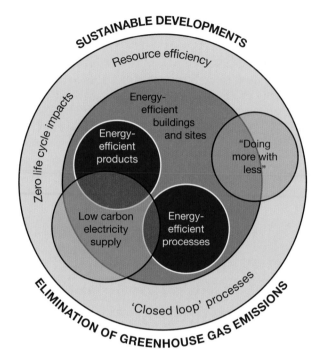

They are in the vanguard of society's transition to a more sustainable equilibrium with the planet's ecology, and play a major role in educating their colleagues. Awareness of the need to reduce carbon emissions began with the enthusiastic few and, as the imperative to meet national legal targets of reducing overall emissions to ever lower levels bites harder and harder, this awareness will continue to spread to every head of the population. Energy managers are catalysts for this change. Inevitably they find themselves alternately enthused and frustrated by the behaviour of people in their buildings, depending on whether they themselves are motivated to join them in battle, or whether, through ignorance, inability or perversity, they form a hindrance to progress.

Energy efficiency is a subset of resource efficiency. For example, reusing waste materials saves the purchase of new ones. Redesigning products, or the production line, so that fewer resources are used, will have the same effect. Both result in the use of less energy. Similarly, water efficiency will not only result in lower water and sewerage bills, but it uses less energy in pumping or heating, producing double wins.

The energy manager has to think holistically; to be able to see the big picture, and how energy, materials and people flow through the system in relation to the demands and conditions imposed upon them. Everyone and everything responds to environmental stimuli and in turn has an effect upon it. All of this can be quantified, and must be, in order to provide an evidence base that justifies the action plan to be implemented.

The need for energy managers

For most top corporations, measuring sustainability, of which energy use and therefore carbon emissions form a great part, has become a way of measuring the quality of management of an organisation; that is to say, the viability of its operations beyond the profit and loss account. For a business to be truly sustainable it must totally transform the way it works, with its customers, investors, its employees and its supply chains. And for these companies, managing their energy efficiently is the aspect of sustainability which attracts the greatest attention.

This effect has yet to filter down to many smaller companies and organisations. Nevertheless, for a management board to have appointed a position of energy manager signifies that they have acknowledged the importance of sustainable energy use within their organisation. Then there are the tens of thousands

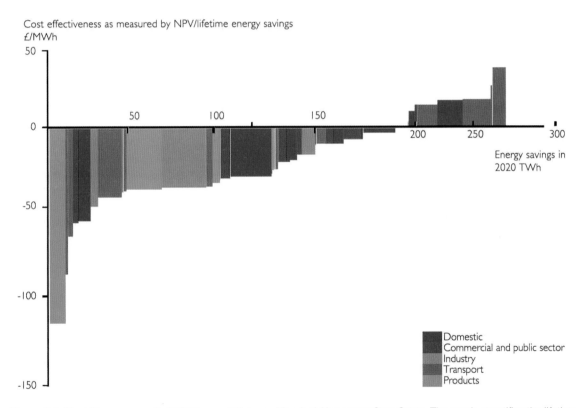

Figure 0.2 The UK government's 2020 Energy Efficiency Marginal Abatement Cost Curve. The graph quantifies the lifetime cost benefits of various energy efficiency measures across different sectors, and is discussed in more detail in Chapter 10. The y-axis represents the cost-effectiveness of a measure, each of which is represented by an individual coloured bar. Any measure which costs more than it saves over its lifetime is represented by a bar which goes over the horizontal axis. The overall message is that the vast majority save money over their lifetime. The net present values are calculated in 2012 terms. The EE-MACC is based on an estimate of the feasible roll-out of energy efficiency measures and takes into account supply constraints for energy-efficient products, only including technology that is already available in the market.

Source: UK Energy Efficiency Strategy, November 2012

of building managers and facility managers, only part of whose responsibilities includes being responsible for energy management. With their labour, their employer often saves a considerable amount of money, more than enough to pay their salary, and reduces the risk of exposure to volatile energy price increases. But it is not just money they save, although that may be their employer's primary motivation. They are also saving carbon, which is increasingly a quantified activity featuring in company annual reports, and as such doing their bit to challenge the advance of global warming and promote the good reputation of the company for sustainable housekeeping.

Legal requirements

There are also numerous statutory requirements motivating the saving of energy. The newly published British Energy Efficiency Strategy looks to achieve 196 TWh of energy savings in 2020, with a reduction of around 11 per cent over the business-as-usual baseline, and a reduction in carbon emissions of 41 $MtCO_2$. The Energy Management Alliance, a forum for the UK's energy management companies and industry bodies, foresees a huge growth in the sector as a result.

The EU's Energy Efficiency Directive has a target of 20 per cent energy savings for the EU as a whole by 2020. It mandates energy audits and energy management by large firms, and stipulates that 3 per cent of public buildings that are owned and occupied by central government must be renovated every year. The recast EU's Energy Performance of Buildings Directive (EPBD) was transposed into national legislation in 2012. Member States are required to set energy use at cost-optimal level, and be measured for a whole system (such as a heating system) rather than at a product level (such as a boiler). This will have to be proven by the installer or designer. Energy performance standards are set for new buildings and benchmarks for existing buildings. 'Consequential improvements' are required to the energy efficiency of buildings undergoing refurbishment, and all buildings must have an Energy Performance Certificate (EPC) available when offered for sale or rent. A small number of buildings are exempt (e.g. some heritage buildings). The EPCs of large buildings to which the public has access must be displayed.

In the USA, there is no nationwide law governing the energy efficiency of existing buildings. Little has been done in this sector and there is huge potential for savings, despite the encouragement of the Energy Independence and Security Act of 2007 (EISA), and the American Recovery and Reinvestment Act of 2009. These have provided finance for improvements; for instance, under the Energy Efficiency and Conservation Block Grant (EECBG) Program. The building sector is the largest consumer of energy in the United States, at around 41 per cent of total US energy use; the industrial sector is also responsible for 20 per cent of energy use.

The LEED (Leadership in Energy and Environmental Design) Green Building Rating System is a voluntary standard for sustainable buildings. LEED includes a standard of measurement for defining a 'green building', and achieving LEED certification is a means of recognising environmental leadership in the building industry and raising awareness of the benefits of environmental building. It is based on well-founded scientific standards and incorporates sustainable site development, water savings, energy efficiency, materials selection and indoor environmental quality. Mandatory Residential and Commercial Energy Conservation Ordinances (RECOs and CECOs) have been implemented by a handful of municipalities as a way to bring the existing building stock closer in line with the energy code requirements for newer buildings.

However, in 2009, President Obama mandated federal agencies to make significant reductions in energy consumption, hoping that the government would 'lead by example' by upgrading many of its facilities. Two years later, the administration tried to jump-start that work by setting a goal for federal agencies to enter into at least US$2 billion of energy efficiency projects within two years. In President Obama's second term, this trend is likely to be accelerated.

Barriers to energy efficiency

If energy efficiency is such a good idea, why is it not practised more widely? The UK's Energy Efficiency Strategy has identified several barriers:

1 Misaligned financial incentives: the person responsible for making energy efficiency improvements is not always the one who will receive the benefits of these actions.
2 Lack of management buy-in: boards may think that energy lacks strategic importance in comparison to other imperatives, especially if energy costs are a small proportion of overall business costs.
3 Hassle costs: perceived disruption caused by making the improvements; for example, building works or production lines halted.
4 Lack of awareness: many people are unaware of just how much can be saved by adopting even simple measures. There is a lack of access to trusted and appropriate information, especially at key decision-making times. Even when present, information may only be generic and not specific and tailored to the situation.
5 Lack of supply: the energy efficiency market itself is underdeveloped, with a supply chain that is still gaining maturity in some areas.
6 Lack of financial support: often financiers fail to appreciate the benefits of investment in energy efficiency, especially if the financial argument is complex. Companies are often reluctant to invest in energy efficiency, seeking short payback times, even if a project is cost-effective at usual interest rates, or on a life cycle basis.

Energy security and procurement

From the point of view of an energy manager, energy security implies securing a constant supply at a realistic price. This is normally achieved via a procurement strategy, but increasingly energy managers are also implementing their own generation schemes, typically from one or more renewable energy or combined heat and power sources. A typical procurement contract will be selected on the basis of competitive tendering. It is common for the winning contractor to be a large group purchasing on behalf of many customers in order to secure an improved price. Agreements come in two forms: fixed and flexible. They will include a variety of energy service and tariff agreements. The contract most appropriate for an organisation depends on individual requirements such as energy use patterns and budget considerations.

Larger companies typically secure the services of an energy procurement consultant whose responsibility is to protect the organisation from risk. This is particularly crucial in a deregulated market. It is important to choose a broker who is competent in reducing the organisation's carbon footprint. Although many will claim this skill, evidence of competence must be forthcoming. One of the easiest ways for an organisation to slash its carbon footprint is by a green electricity purchasing contract which guarantees a supply of renewable electricity.

There are two types of such contract: in the first, the utility guarantees to purchase an amount of renewable electricity from a supplier equivalent to that supplied. In the second, the electricity supplied actually comes directly from a renewable source such as a windfarm, waste-to-energy plant or landfill gas-fired turbine. The latter is to be preferred, although it may be more costly, since it provides more of a market stimulus to invest in new renewable energy generation plant.

Carbon accounting

All large organisations and public bodies in Europe have to account for all the carbon which their buildings emit; this is not the case in the USA. The first stage in implementing carbon accounting is to measure the current or baseline carbon emissions. Organisations then set targets for reduction of emissions, implement systems to monitor them and conduct periodic audits. The energy manager then has to report, both internally and externally, on the progress of the reduction programme against the targets.

Normally, the emissions of a building will only be part of the overall carbon emissions of an organisation, which will, for example, include emissions from transport that are usually outside the brief of the energy manager. The energy manager may therefore find themselves working with colleagues to compile the total emissions for the organisation, in which case their work is merely to feed them summary data when required.

In smaller organisations, or organisations where the only emissions are attributable to those from buildings, the energy manager will occupy both roles. In addition, they may have to motivate staff and perhaps volunteers to work towards the emissions targets.

Aside from transport, there are three main uses of energy which can result in carbon emissions: heating, equipment and machinery. In certain sectors there are also processing emissions from, for example, cement, aluminium or waste processing; and fugitive emissions, for example, from air-conditioning and refrigeration leaks, or methane leaks from pipelines. These are all forms of direct emissions.

Minimise demand, maximise bottom line

The energy manager needs to embody multiple skills as he or she tries to obtain the optimum results within budgetary constraints. Their job is not merely about quantifying the amount of energy used by a building but making a business case for the return on investment of various energy-saving measures. They also need good people skills. This means they have to have an understanding of business planning. Energy, labour and cash flow through an organisation in the same way that food and oxygen and water flow through a living organism. The most efficient use of all of these in combination is the goal of a survival strategy. The creation of a long-term energy management strategy and the communication of that strategy to the board of directors or to one's peers are separate skills but they need to be combined in one individual. If he or she doesn't have presentation skills or powers of persuasion they need to seek help, and this is often available from state-sponsored organisations such as the Carbon Trust in the UK.

The principal philosophy behind designing the strategy is therefore eco-minimalism: doing more with less; obtaining the best results for the minimum investment. Often, this is not investing in a new piece of generation kit (although it might be), it is in investing in better insulation and draughtproofing, since it is always cheaper to save energy than to generate it. The first step then is to look for ways to reduce energy demand. The second is to find the best way to satisfy it. Each way will have its own costs and its own paybacks which will need to be estimated separately and combined in the business plan spreadsheet. Only in this way can individual measures be prioritised for action.

Typical priorities

1　behaviour change – switch off, turn down;
2　draughtproof – remove leaks;
3　insulate to as high a standard as possible;
4　double- or triple-glaze;
5　eliminate thermal bridges;
6　make as airtight as possible;
7　install passive stack ventilation with night cooling or, if not possible, mechanical ventilation with heat recovery;
8　supply the remaining energy renewably only where appropriate.

Standards

This book recommends obtaining the professional qualification associated with energy management: ISO 50001, or any of the other standards (see Table 0.1). It is intended to be read alongside the process of training to attain such certification.

Table 0.1 Standards in energy management internationally

Country	Standard	Description
Worldwide	ISO 50001:2011	The only international framework for industrial plants, commercial facilities or entire organisations to manage energy, including all aspects of procurement and use. Takes account of many of the schemes below.
EU	EN 16001:2009	Covers requirements of energy management systems, now obsolete and replaced by the above.
USA	ANSI/MSE 2000	Voluntary standard for an energy management system. Covers the elements required to ensure continual improvement, sustain savings from energy projects and a strategic energy management plan.
Europe	EN 16247	Defines the attributes of a good-quality energy audit; mandated by the European Commission for Energy Efficiency Directive.
Australasia	AS/NZS 3598-2000	An Energy Audit Standard that represents good practice for energy auditing.
Denmark	DS 2403	Sets out requirements for an energy management system (2008).
Ireland	IS 393:2005	Energy Management Systems Standard to help organisations integrate energy management into their business structures. Shares common management system principles with the Environmental Management System Standard ISO 14001 (superseded).

EN standards apply to the whole of Europe. ISO standards are global. A new standard for energy audits, ISO 50002, and a complete set of EN 16247, will be available by early 2014. These represent an improvement on the current standard, EN 16247-1. EN 16247 is broken down into the following: EN 16247-1 (general); EN 16247-2 (building); EN 16247-3 (process); EN 16247-4 (transport), and EN 16247-5 (qualifications of energy auditor). However, there are differences between EN 16247-1 and ISO 50002. In principle, EN 16247 is focused on energy efficiency, whereas ISO 50002 is aligned to ISO 50001; in other words, it takes the more holistic view of energy performance, defined as energy use, energy consumption and energy efficiency. At some point in the future, Europe will need to decide to withdraw EN 16247-1 and realign EN 16247-2, -3, -4 and -5 to ISO 50002.

Of all the published ISO standards, over 155 relate to energy efficiency and renewables, with many more in development. They cover both generic subjects such as energy management and energy savings, as well as sector-specific solutions for buildings, IT and household appliances, industrial processes and transport, among others. ISO standards for renewables tackle subjects such as bioenergy,

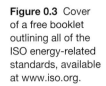

Figure 0.3 Cover of a free booklet outlining all of the ISO energy-related standards, available at www.iso.org.

Source: ISO

ISO&energy
Working for a cleaner, sustainable future

Energy efficiency and renewables are key to meeting the world's energy demand and reducing up to 40% of carbon emissions by 2050

ISO standards represent consensus on concrete solutions and best practice for energy efficiency and renewables

ISO standards open up markets for innovations that address the energy challenge

biofuels and solar power. To develop them, ISO, the International Standards Organization, works closely with key organisations in the energy field, such as the International Energy Agency (IEA), the International Electrotechnical Commission (IEC), the World Energy Council (WEC) and the Efficiency Valuation Organization (EVO), as well as sectoral organisations like the International Commission on Illumination (CIE).

Basic concepts, terms and definitions

The vocabulary of an energy manager revolves around kilowatt-hours, kilograms of carbon dioxide, U-values (R-values in the USA), square and cubic metres, and units of currency. Basic terms and definitions, with other conversion factors and so on, are given in the Appendix at the end of the book.

The scope of this book

This book takes the reader on the journey from novice to expert. Chapter 1 deals with conducting an energy audit, the basis from which all energy management starts. Having discussed how to establish a baseline of energy consumption, using records and metering, it then moves on to the subject of building energy management systems, both wired and wireless, before looking at how to interrogate these to find areas where improvements can be made, including anomaly detection and proactive systems that seek out problems in advance. Chapter 2 is about metering, touching on measurement and verification (M&V) of energy use and how to formulate an M&V plan to reduce that use. The remainder of the book, in effect, explores how details of this plan could be tackled, by order of topic, in a wide range of situations.

Chapter 3 covers the often overlooked issue of how to achieve the cooperation of the rest of the organisation in which the energy manager works, beginning with an organisational energy management appraisal matrix. It is vital to allocate responsibility for areas of energy efficiency for the M&V plan to be effected, and for senior management to buy into the plan. The eight stages of the energy management action plan are then explored.

The next three-and-a-half chapters then discuss technical ways of minimising energy demand, which is usually more cost-effective than investing in new renewable energy generation plant.

This process starts in Chapter 4 with the building envelope, beginning with quick wins in terms of minimising unwanted heat losses or heat gains, and moving on to strategies for improving the airtightness of buildings, leading up to a discussion of the Passivhaus standard. Chapter 5 examines lighting, beginning with ways to maximise the use of daylighting and then looking at the lowest energy-consuming lighting options and controls for different applications. Chapter 6 looks at ways of heating and cooling buildings without an ongoing energy cost, by using passive cooling and passive heating techniques such as solar chimneys and phase change materials, before going on to take a first look at heat recovery, in this case from common appliances such as boilers (UK) or furnaces (USA) and refrigerators. This topic is returned to in more detail in Chapter 7. Chapter 7 tackles the subject of heating and cooling head-on, first introducing the concept of degree days, then looking at sources of renewable heating and cooling by technology, including solar and CHP, moving on to integrated heating, ventilation and air-conditioning systems, finishing on thermal destratification.

Chapter 8 is devoted to minimising water use, which frequently comes within the energy manager's remit because of the energy costs associated with water use. It includes a look at rainwater harvesting and greywater recycling. Once all ways of minimising energy demand have been exhausted, Chapter 9 then examines the various renewable electricity technologies: hydroelectric, renewables-fired cogeneration (combined heat and power, or CHP), wind power, solar photovoltaics, hydrogen fuel cells and other forms of storage.

Before any of the above can be implemented, however, senior management needs to be persuaded of the financial and other benefits of each measure. Thus Chapter 10 presents the financial tools that can enable an energy manager to make a convincing case, but also to compare the relative merits of different projects, in terms of energy, cash and carbon paybacks. Finally, the Conclusion reveals some perhaps heretical truths about what kinds of measures may really be worth implementing, providing a conceptual framework for making investment choices, and urging that sometimes one can achieve more by simply doing less.

Energy-efficient transportation

We will make brief reference to this, since it could come within the remit of an energy manager where an organisation operates fleets of lorries, vans or cars and wishes to lower its entire greenhouse gas emissions. Fleet purchases represent 50 per cent on average of all new vehicle purchases in many countries. Generous tax breaks are often allowed for the purchase of low-emission vehicles, and they are cheaper to run, representing a double win. Hybrid or electric vehicles could be favoured, where the electricity is supplied by renewable energy.

The further an electric vehicle travels each day, the more cost-effective they are to their owners, as long as they are frequently charged. A study by the International Transport Forum[2] found that electric passenger cars currently cost €4,000 to €5,000 more to their owners than an equivalent fossil fuel car over the vehicle's lifetime, but because it will travel greater distances, an electric delivery van costs €4,000 less to its owners over its lifetime than a similar fossil fuel van.

Forklift trucks represent a particular opportunity for changing to fuel cell-driven vehicles, when replacing battery-powered vehicles, because of the reduced charging time and longer range. Marks & Spencer, Coca-Cola, Walmart and FedEx have converted forklift trucks in their warehouses to fuel cells because they keep going for many times longer than battery-powered trucks. The fuel cell stack manufacturers have designed the fuel cells to be exactly the same size as the battery packs they are replacing, so the trucks can be adapted quickly and simply. Customer reaction has been extremely positive.

The big challenge

Energy efficiency is interesting because it can be justified in two ways. Even if climate change wasn't a problem it would be worth saving energy in order to save money. In addition, since climate change is real and one of the biggest challenges facing life on earth, even if it wasn't worth doing to save money it would be worth doing to save carbon emissions. By following all the suggestions in this book, it would be possible to adapt a building or industrial site so that it produces no greenhouse gas emissions at all, and perhaps even sequesters carbon from the atmosphere. It is a huge challenge facing humanity, to reduce and reverse the rate at which greenhouse gases are being emitted into the atmosphere. We hope that this book works as a useful manual or tool to this end for energy managers and students of the subject everywhere.

Notes

1 UK Energy Efficiency Strategy, Department of Energy and Climate Change, November 2012.
2 Smart Grids and Electric Vehicles: Made for Each Other? International Transport Forum, OECD, July 2012 (http://bit.ly/N7bMi4).

1

Measuring energy consumption

Energy savings cannot be made on a sustainable basis without an appreciation of where energy is being used, for what reason, and what may be influencing consumption patterns. Energy use and energy savings must also always be quantified and recorded in order to check that they are accurate. This chapter moves from a basic discussion of how energy is used, through conducting an energy audit, to techniques for regular energy measurement and monitoring. All of this is a prelude to identifying excessive energy use, consequent potential energy conservation measures, and a full measurement and verification (M&V) plan.

Energy units

Energy is described using several different units, but they must be harmonised for comparison purposes. The SI unit of energy is the joule. Other units of energy include the kilowatt-hour (kWh) and the British thermal unit (Btu). One Btu is equivalent to about 1055 joules, and one kWh is equivalent to exactly 3.6 million joules. Full conversion tables are given in the Appendix, together with tables converting fossil fuel energy use into greenhouse gas emissions and other vital information. These are necessary to convert energy use in different contexts to the same units for calculations.

Energy sources

Energy arrives in a building from many directions: from direct sunshine, from an electricity supply, from fuels such as gas and oil which are burned for heating, and from heat arising from activities, including the heat of human bodies. Energy is used in many ways in buildings: for space heating and cooling, for water heating and cooking, and for lighting and equipment such as motors, fans, PCs and freezers/fridges.

Energy efficiency

Energy never disappears: once used, it just becomes converted into another form, usually heat. In this case it is absorbed into the fabric of the building, through it and into the environment outside. It is a general rule that in any system, energy

Figure 1.1 This diagram shows the amount and applications of energy in a block of flats, and how it is satisfied. Red is heat (fuel) and blue is electricity. In this particular case, some heat is supplied by solar water panels and some electricity by solar electric modules, indicated on the right.

Source: Premnath Sundharam

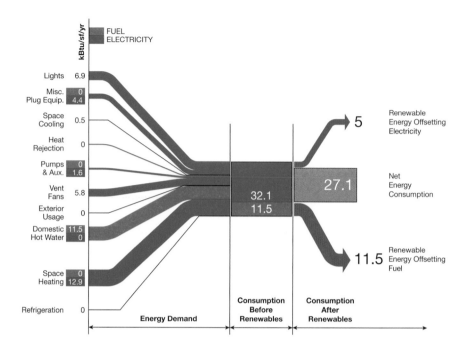

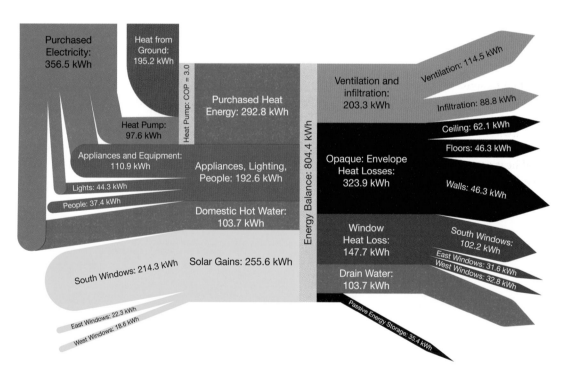

Figure 1.2 Diagam showing the energy balance of a hypothetical building. Sources of energy entering are on the left, and ways in which it leaves are on the right. The total energy balance is quantified in the centre.

Source: 'Preliminary Investigation of the Use of Sankey Diagrams to Enhance Building Performance Simulation-Supported Design', William (Liam) O'Brien of Carleton University, Ottawa

in equals energy out. But some of this energy may be stored within the building fabric, or even in specific energy storage such as hydrogen, hot water tanks or batteries. The control of heat energy in and out is governed by the building structure and fabric itself, how this is managed by the people within it, and the heating, ventilation and air conditioning (HVAC) system. These issues are covered in Chapters 3 and 6.

Efficiency is measured by the amount of useful work conducted by the energy supplied. This is where energy is converted into power. The energy which is wasted in this process appears as heat. Some heat may be reclaimed and reused, as we shall see in Chapter 6, but there is a natural limit to this process. Sometimes this heat is unwanted, but it must still be efficiently dealt with; an extreme case of this is in data centres or server rooms, which are discussed in our companion book, *Energy Management in Industry*.

Specifically, the following are the definitions of efficiency for different types of energy use:

- Electrical: useful power output per electrical power consumed;
- Mechanical: the proportion of one form of mechanical energy (e.g. potential energy of water) that is converted to mechanical energy (work);
- Thermal or fuel: the useful heat and/or work output per unit of fuel or energy consumed;
- Lighting: the proportion of the emitted electromagnetic radiation usable for human vision;
- Total efficiency: the useful electric power and heat output per fuel energy consumed.

It often makes sense to talk about the efficiency of an entire system or process, such as how much electricity is consumed to produce 1000 widgets. It is extremely important to be clear on the units that are being used and the type of efficiency being measured, otherwise confusion can result.

For example, there is a difference between Europe and America in the definition of heating value. In the US and elsewhere, the usable energy content of fuel is typically calculated using the higher heating value (HHV), which includes the latent heat for condensing the water vapour emitted by burning the fuel. In

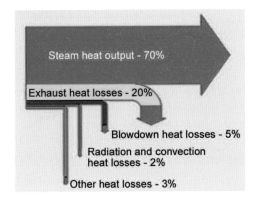

Figure 1.3 Diagram showing the heat losses from a non-condensing boiler. All processes lose energy as heat.

Source: Author

Europe, it is the convention to use the lower heating value (LHV) of that fuel, which assumes that the water vapour remains gaseous, and is not condensed to liquid water so releasing its latent heat. Using the LHV, a condensing boiler, which makes use of this latent heat, can achieve a "heating efficiency" in excess of 100 per cent, whereas, of course, using HHV, this efficiency is around 90 per cent, compared to 70 to 80 per cent for non-condensing boilers.

The aim of energy management is to improve the overall efficiency of the entire systems for which the manager is responsible: to get more work out for the energy in, and certainly to make use of the free energy that is available, principally from the sun. It is important to note that this free energy, most of which is called 'passive gains', is not directly measured, yet it is a powerful factor within the controlled environment. In calculations to determine building efficiency strategies, it needs to be taken into account. This is covered in Chapters 3 and 5.

Energy audits

A European and world standard for energy audits, published in October 2012, BS EN 16247-1, explains the process of conducting an energy audit in great detail, defining the attributes of a good-quality energy audit, from clarifying the best approach in terms of scope, aims and thoroughness to ensuring clarity and transparency. It specifies the requirements, common methodology and deliverables for energy audits. It applies to all forms of establishments and organisations, all forms of energy and uses of energy, excluding individual private dwellings, and is appropriate to all organisations regardless of size or industry sector.

It was developed by members of ESTA (the Energy Services and Technology Association), the Energy Institute, Institute of Chemical Engineers, and Energy Services and Technology Association in Britain, in response to the 2006 EU Directive and in anticipation of the Energy Efficiency Directive, that mandates member countries to create regular energy audits for large organisations. It complements the energy management system standard, ISO 50001:2011 which identifies the need for clear energy auditing. At the time of writing, further, more specific standards are being developed but will be available soon: energy audits for Buildings (EN 16247-2), Processes (EN 16247-3) and Transport (EN 16247-4).

The first step is to discover where all the energy comes from and where it goes. When this is done it can be compared to an established benchmark of energy use for the building, company, organisation or facility.

Establishing the baseline

Energy used is either metered or unmetered. Metered information may be gathered from invoices from suppliers and utility companies and from the meters directly. It is necessary to make a list of all meters and sub-meters. This will detail their location, what fuel or energy source they measure, the zone or equipment they serve, the units of measurement, the reading recorded at present (with the time and date), and some other identification such as a serial number. Often, they are photographed. A policy should be set to determine how frequently these are read and by whom. These people need to be taught how to do it, so that everyone is reading the meter in the same way.

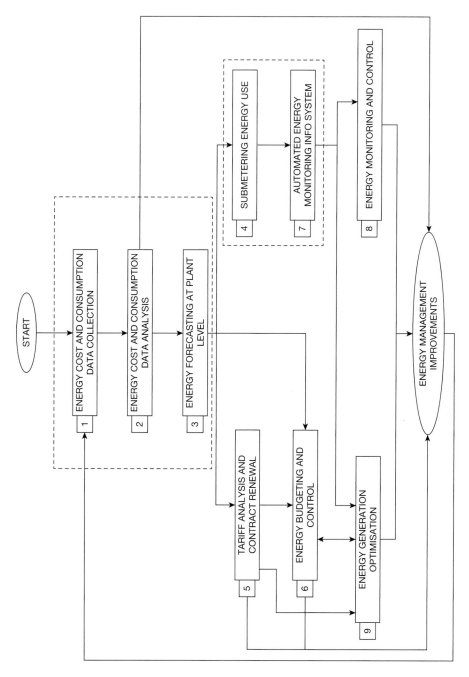

Figure 1.4 The process of auditing, benchmarking and performance optimising.

Source: Carbon Trust

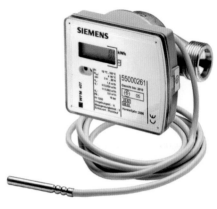

Figure 1.5 Modern smart meters provide half-hourly measurements, and even enable the utility companies to provide their customers with cheaper electricity at certain times.

Source: Siemens press picture

Figure 1.6 Modern heat meters can measure the cumulative heat consumption since starting up the heat meter, and cumulative water consumption up to the last due date, and the current return temperature.

Source: Siemens

Fuel that is used for heating and transport will be recorded on invoices. This will include oil, solid fuels, biomass, petrol, etc. Invoices may not be accurate, and it may not be possible to allocate particular purchases to the time period over which that fuel is consumed. Invoices may be duplicated or fuel not delivered but billed for. More accurate information may be obtained from the use of heat meters (e.g. for solar water heating, otherwise unmetered), fuel use measurement, and, in the case of transport, asking drivers to record their fuel purchases and use, mileage, etc.

Having made an assessment of the sources of information for an audit, the energy manager must then come to a decision about the best periods over which to record that information: monthly, weekly, daily or hourly. This will vary from installation to installation. The ideal is that the reporting periods are the same for each type of fuel and power usage. This facilitates and optimises subsequent calculations and the presentation of data in graphical form. Most stakeholders will at the very least want to see annual information, and preferably quarterly and monthly figures. Company directors may wish presentational information to coincide with the financial year used by the company for annual reports.

For example, in the case of fuel use, measurements of stock levels should be taken at the beginning of each week, if that is also the period over which electricity is to be measured. The amount of fuel use is then the starting amount, plus any amount delivered during that period, less the finishing amount. From the point of view of analysis, weekly, daily, hourly and, preferably, half-hourly information about heating and power requirements is needed, especially for comparing one year with another and one period with another, since months contain an odd number of weeks.

Summary of sources of fuel information

- Utility bills:

  - gas;
  - electricity.

- Meter readings:

  - gas;
  - electricity;
  - heat;
  - fuel.

- Transport:

  - driver records and receipts, mileage.

- Fuel:

  - invoices;
  - stock level measurements (oil, LPG, wood pellets, coal, etc.).

This information needs to be recorded for each process, zone, building, piece of equipment or vehicle that is being measured. The same recording period (weekly, daily, hourly) needs to be used for each data set. All records are entered into a database. Units need to be converted so that they are compatible (see Appendix for conversion figures). An additional column can be made in the database to convert these figures into tonnes of carbon-dioxide-equivalent emissions. A final column, for a building, would include the floor area in square metres or square feet. This would enable the total energy use to be divided by the total area to give a figure per unit of area. This would enable it to be compared to other similar buildings. The spreadsheet also enables many other calculations and summary reports to be made, graphs and trends to be seen.

Other information that may be recorded includes:

- building occupancy;
- process ongoing;
- outside temperature;
- light levels;
- degree days (see Chapter 7).

Recording the outside air temperature and light levels permits the correlation of energy use with outside conditions and will enable predictions of future energy use based on time of year and weather conditions. Monitoring of interior light levels may also be installed and linked to lighting systems to enable the provision of the optimum level of lighting depending on ambient conditions. This will be considered in more detail in Chapter 4.

Independent rating systems

It is possible to obtain third-party audits, which can be helpful in giving confidence and independent verification. In the UK, rating of the energy performance of many public and commercial buildings is compulsory and conducted by independently certified assessors. Depending on the type of building, they will be Display Energy Certificates or Energy Performance Certificates. The assessment results in the issue of certification are described in figures 1.7 and 1.8. DECs must be publicly displayed and available on a website so that building managers may compare the performance of their building against others. Clients also receive a summary of measures which may be taken to improve the rating.

In the USA, there is no compulsory system for rating buildings. However, the Environmental Protection Agency's ENERGY STAR Program has developed two rating systems for several commercial and institutional building types and manufacturing facilities: the ENERGY STAR Portfolio Manager/Benchmarking Program; and the ENERGY STAR Buildings Five-Stage Approach Program, which is essentially a measurement and verification (M&V) plan, although it culminates in a certificate.

The Portfolio Manager rating is calculated based on the information a client enters about their building, such as its size, location, number of occupants, number of PCs, etc. The system estimates how much energy the building would use if it were the best performing, the worst performing, and every level in between. The system then compares the actual energy data entered to the estimate, to determine where the building ranks relative to its peers. These ratings, on a scale of 1 to 100, help to benchmark the energy efficiency of specific buildings against the energy performance of similar ones; a rating of 50 indicates average energy performance, while a rating of 75 or better indicates top performance. Buildings with benchmark scores of 75 or higher are eligible for the ENERGY STAR label for buildings, which can be displayed to convey performance excellence to tenants, customers, and other occupants. The ratings are used by building and energy managers to evaluate the energy performance of existing buildings. All the calculations are based on source (primary) energy. The use of source energy is the most equitable way to compare building energy performance, and also correlates best with environmental impact and energy cost.

The ENERGY STAR Buildings Five-Stage Approach for energy efficiency looks at a whole building's systems to attempt to achieve average energy savings of 30 per cent for the whole building. The five stages are:

1 Green Lights: installing energy-efficient lighting systems and controls that can provide substantial energy savings at low cost.
2 Building Tune-up: performing a comprehensive tune-up of the entire facility to get it into peak condition.
3 Other Load Reductions: finding other opportunities for increasing a building's energy efficiency such as purchasing ENERGY STAR office equipment, installing window films, and adding insulation or a reflective coating to the roof.

4   Fan System Upgrades: rightsizing fan systems, adding variable-speed drives, or converting to a variable-air-volume system, if appropriate.
5   Heating and Cooling System Upgrades: replacing chlorofluorocarbon chillers with small, more energy-efficient models to meet the building's reduced cooling loads and upgrading other central plant systems.

There are two more advanced American standards: ASHRAE 189.1 for new buildings and LEED. Their "better-than-code" approach entails comparing a building's modelled energy use to the modelled energy use of the building if it just meets the minimum requirements of the building energy code. EPA's ENERGY STAR energy performance scale is based on the measured performance of buildings in the market, rather than code specifications or modelled energy use.

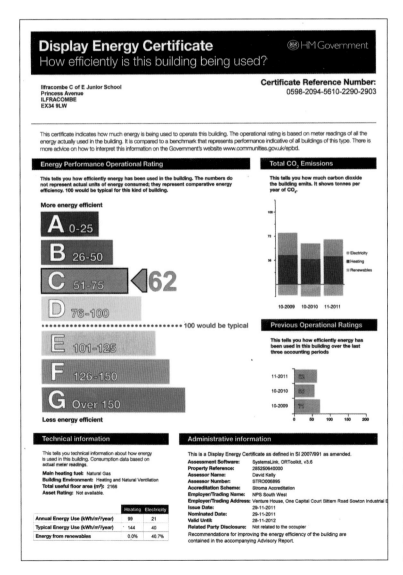

Figure 1.7 Display Energy Certificates are compulsory for all public buildings in England and Wales with a total useful floor area >1000m² which are frequently visited by large numbers of persons, such as schools and hospitals. Some privately owned buildings that provide services from public funds such as leisure centres, museums and theatres may also need to display a DEC.

Source: Ilfracombe School, Devon, England

The American Society of Heating, Refrigerating and Air-Conditioning Engineers (ASHRAE) energy labelling system is called the Building Energy Quotient. It involves independent assessors investigating the site and producing a report on key building performance indicators, which is also benchmarked.

The US Green Building Council's LEED (Leadership in Energy and Environmental Design) certification system also provides third-party verification of new but also existing buildings. Its distinguishing mark is that it addresses the entire life cycle. To earn LEED certification, a project must earn 40 or more points on a 110-point scale. Various rating systems exist to address different types of projects, including health-care facilities, commercial, retail, schools and entire neighbourhoods. Credits are given for:

- ecological siting;
- water efficiency;
- energy performance;
- resource efficiency;
- indoor environmental quality.

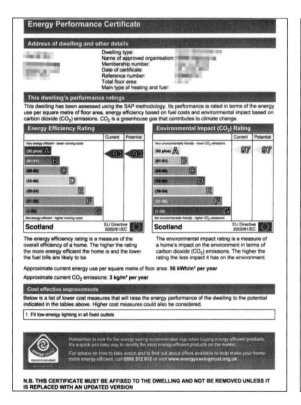

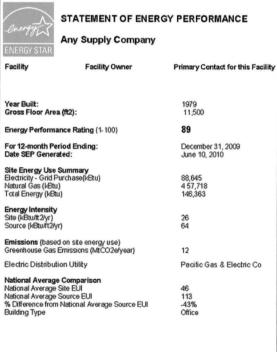

Figure 1.8 Energy Performance Certificates are compulsory for all non-public buildings in the UK. These may contain useful information, but they are just the start of a diligent process of energy management.

Source: Author

Figure 1.9 In the USA, the EPA ENERGY STAR Portfolio Manager rating system is voluntary and applies only to some buildings in America. At the end of the process a certificate like this would be awarded.

Source: EPA

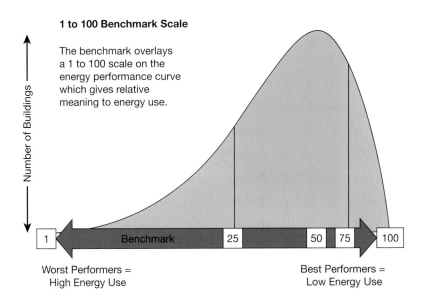

1 to 100 Benchmark Scale

The benchmark overlays a 1 to 100 scale on the energy performance curve which gives relative meaning to energy use.

Number of Buildings

| 1 | Benchmark | 25 | | 50 | 75 | 100 |

Worst Performers = High Energy Use

Best Performers = Low Energy Use

Figure 1.10 This curve shows the skewing towards the positive of the proportion of buildings which get different ratings under the EPA ENERGY STAR Portfolio Manager system for buildings.

Source: EPA

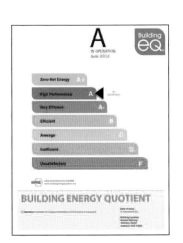

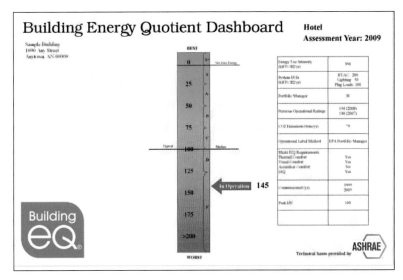

Figure 1.11a and b AHRAE's Building eQ label and dashboard awarded to buildings that volunteer to go through this more stringent independently assessed American rating system.

Source: ASHRAE

An audit establishes the baseline, and benchmarking compares this with other similar buildings. This will help to develop a strategy for the ongoing recording of energy use information. To acomplish this, a metering strategy is needed. This is the topic of the next chapter.

2
Metering

Metering has undergone a revolution in recent years. This chapter looks at the latest technologies, which are increasingly wireless- and cloud-based.

Most metering hardware nowadays is supplied with software that enables the energy manager to interrogate the data supplied and have it presented in visual, graphical form, so that they will be able to see the actual load profiles over 24-hour and weekly periods. Their eagle eye will examine the data to spot unusual and inappropriate uses of energy, for example, to correlate when lighting is switched on during the actual occupancy of premises. Similarly, the heating or air-conditioning profile should reflect occupancy levels at that time, with the thermostats set at the appropriate temperatures. If a space is being lit or heated when it is not occupied, then something may be done about it.

Most businesses which consume over a certain amount of electricity will have meters fitted which record half-hourly usage. If such meters are not present there is a strong case for having them fitted, because only through this route can detailed analysis and savings be made. An alternative is the use of data loggers (see below).

In large installations, for more detailed analysis, or for operations with multiple buildings, sub-metering is employed. Additional meters are installed in places where it is considered beneficial to monitor energy use. Depending on the nature of the company or organisation, this may be hospital wards, schoolrooms, sports halls, chiller rooms, server rooms, office suites, etc. In other words, wherever there is a risk of excessive energy being consumed, that is where a sub-meter should be placed. The degree of "excessive" is a subjective evaluation, based on the "risk of acceptable undetected loss". For example, a section of the building may have manually controlled heating and windows, which could mean that heat is unnecessarily supplied and wasted at times. Whether this loss is considered excessive and worthy of metering may depend on the total amount of electricity being consumed.

Meter types

The following is a non-exhaustive list of possible metering solutions that are available:

Electricity

- Regulator-approved meters.
- Measurement Instrument Directive-compliant meters. The Measurement Instrument Directive (MID) is a European standard which covers many types of measuring instruments. It is important to check whether a meter is compliant with the regulator (in the UK, Ofgem) or MID, or both, for the purpose for which it is intended.
- Combined meters are microprocessor-based energy meters with many different functions, measuring active and reactive energy in both directions of energy flow and able to display instantaneous values such as current, voltage and power.

Electricity meter output options include the following:

- Pulse output: this is a relay contact (more often that not an electronic relay) inside the meter that closes momentarily when the meter advances an increment. This contact is fed into metering data loggers where they are counted and stored.
- MODBUS: here, the data logger asks the meter for its current reading and the meter then responds. Any value may be read remotely. MODBUS is the most common protocol in use; however, there are others such as M-BUS and BACNET. If no response is received from this type of meter the data logger can raise an alarm, whereas if a pulse signal fails this could be seen as meaning zero consumption. Pulse-based systems are limited to logging only kWh or kVArh (kiloVolt-Amps reactive per hour: the difference between working power (measured in kW) and the total power consumed (measured in kVA)), whereas MODBUS-connected systems can log most parameters that a multi-function meter would measure, such as current, voltage, power factor, frequency, instantaneous power as well as the energy register displays.
- Analogue outputs and alarms: these can be programmed to give a 4–20mA signal or similar in sympathy with the amount of instantaneous power being consumed. Others have alarm outputs which may be programmed to signal various conditions.
- Current transformers: these are used when the primary circuit rating of the feeder that is being monitored is too large to be connected directly to a meter. They are available with primary ratings from 1–5000A. They can be solid or split core, three-phase low voltage or summation.

Water meters

There are hot (up to 90°C/194°F) and cold (up to 30°C/86°F) water meters which both come in various types: single jet, multi-jet, ultrasonic, screwed connection and Woltmann Flange fitted. Both may be read remotely by connecting up the output contact to an energy management system or building management system.

Heat meters

A heat meter is used to measure the amount of heat (in Btu or kWh) that has been dissipated in a loop of pipework; for example, feeding a heating or processing system. A heat meter consists of the following:

- The meter, which measures the flow of liquid in the pipework and has a pulse output connected to the heat integrator or calculator.
- Two thermocouples, one in the flow leg of the pipework measuring the hot water entering the loop, the other in the return leg measuring the temperature of the water after it has been through the loop.
- A heat integrator, to calculate the amount of kWh consumed, using the flow and the temperature difference calculated. They have an electronic display and an optional pulse output which could be connected to a building management system or energy management system. It is possible to obtain heat meters with either MODBUS or M-BUS communications to enable intelligent connections to remote systems.

Ultrasonic heat meters remove the need to physically adapt the pipework, enabling much shorter installation times and incorporates flow and heat calculations in one unit, which saves having to commission two separate meters.

Meter data loggers

Meter data loggers are the core of any energy management system. They collate data from a number of meters (water, electricity, gas or heat meters, etc.) in real time, storing it in local memory. Data loggers group these consumption values in time periods, the most common being half-hourly. When viewed as a graph, these values form a profile of the energy consumption for that meter or feeder.

They can take data in the form of basic metering pulses or intelligently via MODBUS on an RS485 network. Some meter data loggers include basic reports that may be accessed via the internet, using secure File Transfer Protocol (FTP) to a browser, intranet or customisable e-mail, transmitted via a local ethernet network or wirelessly. It is possible to access data within the data logger easily and interpret the values in a third-party energy management software package or in Microsoft Excel.

Wireless and energy harvesting meters

Traditional meters use cabling and consume power, making them expensive and time-consuming to install. Wireless energy monitoring systems use the internet, and provide two-way communication. They are less intrusive, far easier to fit, and so can be around 50 per cent cheaper to install than cabled meters. They can be installed in four days, as opposed to eight days for cabled systems. This means that every building or unit can have a programmer, and sensors for temperature, light and human presence. In this way, wireless control of lights, heating and air conditioning, shading, motors, valves and most technologies is possible. This solution can achieve 15 to 30 per cent savings on energy use, and therefore have,

Figure 2.1
No batteries or wires: this sensor, attached to an electric motor, relays information about its status and takes its power from the vibrations of the motor itself.

Source: Siemens press photos

Figure 2.2 The WEMS controller 1 is a mains-powered intelligent wireless input/output module for WEMS International's wireless building management system.

Source: WEMS International

Figure 2.3 The WEMS programmer is an embedded computer system that hosts WEMS International's wireless building energy management application software, along with web server and other services that may be required. For communication with WEMS wireless modules the WEMS USB wireless interface module (ASD03N) is attached.

Source: http://www.wems.co.uk/energy_management_system_products.

on average, a payback period of below 18 months. The latest generation can work with or without batteries, and is largely service free. Systems are modular, allowing for progressive expansion as budgets allow.

With battery-free systems, energy is harvested from anywhere in the surroundings: linear motion, light, and differences in temperature and heat. For example, they can use thermistors to harvest the heat, PV cells to run off light, or piezoelectric crystals to run off pressure changes, such as from vibrations, to send a unique radio message to the receiver module. This modest amount of energy is sufficient to send the wireless signal or turn on a light.

The range of wireless subsystems is up to 300 metres outside and up to 30 metres inside buildings. They communicate in the 868 Mhz or 315 Mhz frequency band at a rate of 125 kb per second. Technical means are used to minimise the possibility of transmission errors. Data security protection is supplied as well as protection against attacks.

Building energy management systems

From such a metering system it is a simple step to a building energy management system (BEMS; WEMS = wireless building energy management system). This is

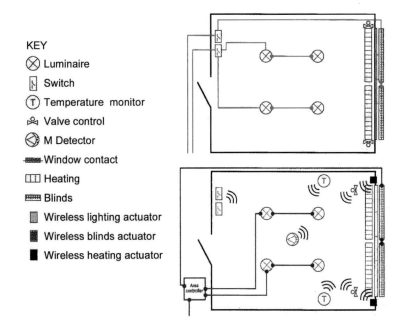

Figure 2.4 A simple wireless energy management system layout for an office. The top diagram shows conventional wiring linking up blinds, valves and heating, lighting and switches. The bottom diagram shows the wireless conversion, with a motion sensor in the middle of the lighting, and two temperature monitors controlling the heating system. The switches are also wireless. Everything is connected to the area controller which also monitors and records activity. A 30 per cent energy saving is claimed, with 70 per cent less cable required.

Source: Author, with thanks to EnOcean

a computer-controlled system that may be used to monitor and control a building's power systems, including lighting, heating, ventilating and air conditioning. Its basic functions include monitoring, controlling and optimising energy usage.

Its aim is to manage the environment within a building so that the use of energy perfectly balances with the use made of it, so as to optimise both. This could involve distributing and managing the level of lighting, heating and ventilating to suit the changing levels of activity within different areas throughout the day, week and year. The power used by machinery, from computers and printers to motors and drives, can be managed.

The level of control may be refined to whatever specification is required, for example, by splitting areas into different zones within, say, an open-plan office, warehouse, large retail unit or factory. Systems are also available that extend the management of resources from energy only to include water, air and steam. So-called WAGES (water, air, gas, electrical, steam) systems are intended for industrial applications. Water is a costly resource that can also consume energy, for example, in pumps and heating. We will look at water management in more detail in Chapter 8.

The supply of sophisticated software for BEMS and M&V is developing fast. It means that the tools available to energy managers are very different from what they were ten years ago. Increasingly, software can be cloud-based and involve modular, downloadable apps for many different purposes. For example, apps are available that keep track of tenants' utilities, monitor buildings' energy performance, benchmark energy consumption, compare usage among other buildings in a portfolio, and display the carbon emissions associated with a building, or group of buildings. Some companies have made their application programming

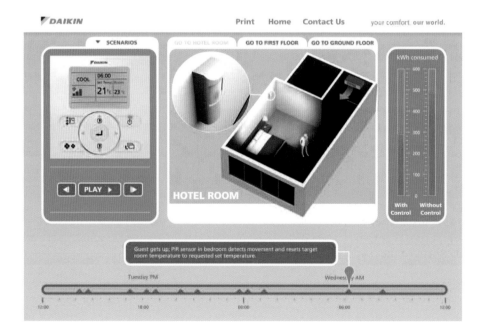

Figure 2.5 The use of occupancy sensors to control air conditioning and temperature in a hotel room via a BEMS.

Source: Daikin

May 04 2012

Outside Temp 21°C
Humidity 53%

Commercial Centre

Building A1

Alarms

23°C	▮	Reception
20°C	▮	Floor 2
21°C	▮	Floor 3
22°C	▮	Meeting room 1
21°C	▮	Meeting room 2

www.trendcontrols.com

interfaces (APIs) available to developers to design apps for the marketplace. Some combine building intelligence and building energy management applications, offering insight for both financial and operational decision-makers.

Third-party data services

In addition, many electricity and gas suppliers provide energy use data services for business customers. This provides access to a dashboard that enables them to compare performance with baselines and prepare reports. Such services are now a standard part of contractual arrangements with suppliers, but they do not all provide the same level of quality service, so it may be worth shopping around. On the other hand, it may be considered that it is not sufficient to rely solely on information supplied by the utility company or energy supplier, on the grounds of objectivity and independence. Objective, independent monitoring provides a level guarantee in terms of risk mitigation for clients and energy service companies. It means that performance analysis can be verified, yielding confidence and transparency for everyone concerned. In all cases, for greenhouse gas reporting purposes software should provide the functionality to report energy used in the form of the corresponding carbon-equivalent emissions.

Figure 2.6 Splitting a building into zones can help target energy use. This is accomplished using software like this, which may be programmed to automatically take into account occupancy levels, outside temperatures and humidity.

Source: Trend Controls

Figure 2.7 Once the baseline is known, specialist software can allow the exploration of scenarios that determine various energy and cost savings depending on measures taken.

Source: BuildingIQ

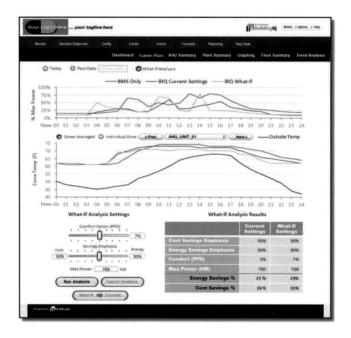

Case study: BT

BT (British Telecom) has installed a wireless building energy management system across 2000 telephone exchanges. These range in size from 400 m² to 17,000 m². BT anticipates saving 22 per cent of its energy use from this measure. Richard Tarboton, BT's Director of Energy and Carbon, says he believes that this initiative 'has underpinned the success of our smart energy programme at BT. It gives us remote control over the air handling systems and enables us to optimise energy usage. As a result of the proven success, we have now agreed a contract to expand the programme further.'

Case study: Farmfoods, UK

Farmfoods is a company that specialises in frozen food; therefore, unsurprisingly, freezers are responsible for most of its electricity consumption at its 300 sites. Following the introduction of a wireless building energy management system, it managed to reduce this consumption by 18 per cent. The return on investment was 17 months.

Anomaly detection

The savings gained in the case studies above are achieved by judicious interrogation of the data provided by these energy management systems. For example, one may notice that heating systems are cutting in an hour earlier than necessary at certain times of the year, and cutting out later than the need to provide an acceptable level of comfort. One may also discover that patterns of heating are being repeated at weekends when no one is in the building. Adjusting the programmer and thermostat to account for this would achieve significant savings at no cost to occupants' comfort.

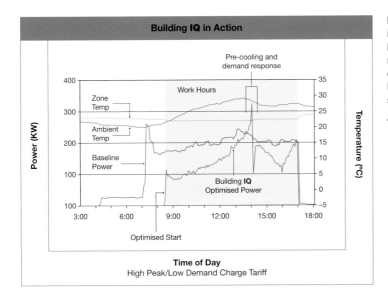

Figure 2.8 In the above example it may be seen that heating was kicking in an hour before it was needed. After optimisation, energy use was matched with building occupancy, achieving significant savings.

Source: BuildingIQ

Figure 2.9 The levels at which heating kicks in can be matched to ambient temperature.

Source: BuildingIQ

Data loggers

It may be felt that, for economic reasons, if most of the savings that can be made by monitoring systems are due to the identification of anomalies, then the same result could be more cost-effectively achieved with a temporary data logger. These work by clamping the device on to a wire and transmitting electricity use data back to the logging device over whatever period the energy manager deems to be useful. Having recorded sufficient information and identified ways in which energy can be saved at this particular location, the device is then removed and positioned somewhere else, where the operation is repeated.

Data loggers may be appropriate in situations where continuous fine-grained data recording is not required, where budgets are low, and in smaller facilities. One would, however, need to keep checking back in the same locations to make sure that the previous high level of energy use has not returned for one reason or another. One may therefore cycle the logger through, for example, ten separate locations over ten weeks to fine-tune the adjustments.

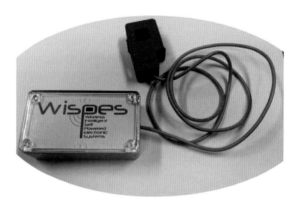

Figure 2.10 A data logger is clamped on to any cable for on-the-spot power consumption measurement.

Source: WISPES

Case study: Scottish Parliament building at Holyrood

The Scottish Parliament Building's energy efficiency credentials have been confirmed by its EPC rating. An Accredited Energy Assessor from IES Consulting undertook the EPC assessment. The energy performance of this highly complex building, which was designed to pre-2000 standards, was analysed using Scottish Building Standards approved Virtual Environment software in order to determine the EPC rating.

Non-domestic buildings are rated for Energy Performance Certificates on predicted energy use calculated using the official UK government's Simplified Building Energy Model (SBEM) or other approved Dynamic Simulation Model (DSM) software. Commercially available building performance modelling tools enable both SBEM and DSM calculations from the same 3D input model. The A to G EPC rating grades the building based on calculated CO_2 emissions as determined from its official energy use. For more information on SBEM and DSM see the UK government Department for Communities and Local Government website: communities.gov.uk.

Figure 2.11 Thermographic image of the Scottish Parliament building composited with a computer-generated image of the same scene.

Source: IES Consulting

Proactive systems

The above systems allow only reactive behaviour on the part of the energy manager. Because weather and energy market conditions are highly dynamic, further efficiency gains can derive from having a real-time ability to modify systems' behaviour. Energy management software is available that can predict and then automatically optimise a building's energy use, continuously adjusting its management system's settings to meet the needs of its occupants and processes each day, while delivering savings.

This approach may be suitable for large concerns that are heavy energy users, who often buy electricity or gas on a daily basis based on fluctuations and flexibility in electricity markets in order to maximise savings. Energy pricing can be volatile. Energy use can depend dynamically on many external factors, not least of which is weather. Utilities and managers therefore often have inaccurate data as to what demand reductions in a given site or building could potentially deliver in terms of savings. The Australian Government Research Organisation (CSIRO) has tested predictive modelling and system control software in this context. It has been shown to produce up to 40 per cent ongoing HVAC energy

Figure 2.12
Predictive modeling can help optimise HVAC operating parameters. This building energy management system allows operators to vary the HVAC system according to building occupants' activities and clothing in relation to temperature and humidity.

Source: BuildingIQ

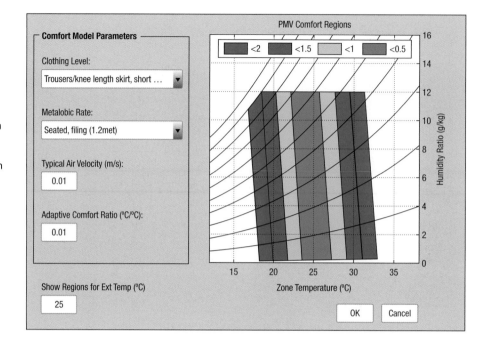

savings and up to 30 per cent peak load reduction during demand reduction trials. Office buildings, retail shopping malls, utility programmes and US Department of Energy facilities are among the sites taking part in these trials. Among the inputs which such sophisticated software requires are energy price signals, weather forecasts and predictive dynamic models of the building. This enables the software to anticipate trends and minimise HVAC requirements.

Such a system would automatically and continuously fine-tune HVAC operations to minimise energy use, connecting a building to the smart grid. It would mean that building owners, energy managers or facility managers are able to access information in real time to meet demand response energy reduction targets, while meeting the requirements of the occupants for comfort. Such innovations still require energy managers to monitor them and ensure that the solution is appropriate in every case, but by automating many tasks it leaves the manager free to engage in active campaigning, publicity and behavioural change activities which are, arguably, the most important part of his or her role.

Case study: The New York Greener, Greater Buildings Plan

The New York Greener, Greater Buildings Plan is part of PlaNYC, an overarching sustainability plan, designed to address energy waste in large existing buildings over 50,000 square feet. These account for almost 45 per cent of the city's greenhouse gas emissions. It was enacted in 2009 with the passage of four local

Figure 2.13 The launch of PlaNYC on a building's green roof.

Source: PlaNYC

laws and the establishment of the New York City Energy Efficiency Corporation (NYCEEC) for financing. It seeks to help decision-makers by giving them information that encourages the pursuit of cost-effective energy efficiency measures (PlaNYC 2011).

These local laws (LL) oblige managers of all buildings over 50,000 square feet:

- LL84: to benchmark their energy performance once a year;
- LL87: to conduct an energy audit and retro-commissioning study every ten years;
- LL88: to upgrade all lighting in commercial space to meet a code standard, and install sub-metering;
- LL85: to adopt a local energy code.

It is estimated that together these measures will reduce greenhouse gas emissions by almost 5 per cent, have a net saving of US$7 billion, and create roughly 17,800 construction-related jobs over ten years. The four laws are coming into effect one by one, and with them have appeared new jobs dedicated to planning and carrying out the efficiency measures. Local Law 84 has involved 3000 public buildings such as libraries, police and fire stations, and schools, and 16,000 private commercial and rented accommodation buildings. Nearly 1.8 billion square feet of built space, which is equal to the built areas of Boston and San Francisco combined, have been benchmarked. This represents the largest collection of benchmarking data gathered for a single jurisdiction.

The benchmarking found that New York City's buildings perform significantly better than the national average, having a median ENERGY STAR score of 64 out of 100 according to EPA comparisons, although the weather-normalised energy use intensity for New York's buildings is comparable to the rest of the Northeast. This suggests that the city's high scores may be attributable in part to the age of the city's building stock, which is similar to the rest of the region.

A key finding was that if all comparatively inefficient large buildings were brought up to the median Energy Use Intensity (EUI) in their category, New York City consumers could reduce energy consumption in large buildings by roughly 18 per cent and GHG emissions by 20 per cent. If all large buildings could improve to the 75th percentile, the theoretical savings potential grows to roughly 31 per cent for energy and 33 per cent for GHG emissions. Since large buildings are responsible for 45 per cent of all citywide carbon emissions, this translates into a citywide GHG emissions reduction of 9 per cent and 15 per cent respectively. Much of this improvement could be achieved very cost-effectively through improved operations and maintenance.

Cities such as Philadelphia, Seattle, Chicago and the District of Columbia are due to enact similar initiatives.

See: www.nyc.gov/ggbp/

The Measurement and Verification (M&V) Plan

We have reached the point now where we can begin to consider a Measurement and Verification (M&V) Plan. An M&V plan is a document which defines project-specific M&V methods and techniques that will be used to determine savings resulting from specific energy efficiency projects. The plan may include either a single option that addresses all the measures installed at a single facility, or several M&V options to address multiple measures installed at the facility.

If a company is ISO 50001 certified, or aiming for certification, then it will be required to implement and maintain a plan of this nature. A well-organised library of documents containing the necessary levels of detail is vital for successful compliance with this standard. In addition to providing accurate and conservative methods to calculate energy savings, a good M&V Plan is clear, consistent and repeatable. It is understandable by anybody else and may be given to subcontractors or an Energy Services Company (ESCo); for example, in the form of a long-term contract. It is therefore very important to ensure that all assumptions, procedures and data are recorded properly so that they may be easily referenced and verified by others.

M&V activities include site surveys, energy measurements, metering of key variables, data analyses, calculations, quality assurance procedures and reporting. Sample specifications are provided by the US Department of Energy Federal Energy Management Program and the Carbon Trust in the UK. In general, the contents of a project-specific M&V Plan should provide an overview of the energy conservation measure and verification activities, including the following:

- the goals and objectives, with key performance indicators, such as the energy, cost and carbon emission savings expected;
- the scope and nature of the work;
- the techniques to be used for each measure;
- the key physical characteristics of the facility, system and measure;
- the critical factors that affect energy consumption of the system or measure;
- the key baseline performance characteristics such as lighting intensities and temperatures;
- the baseline operating conditions such as loads and hours of operation;
- all measurements, data analyses procedures, algorithms and assumptions;
- all performance period verification activities, including the parameters to be measured, period of metering, accuracy requirements, calibration procedures, metering protocols and sampling protocols;
- the schedule for reports and procedures.

The plan will describe the source of all savings, including energy, water, O&M, and other (if applicable). There will be details of the baseline data collected, including the following:

- the parameters monitored/measured;
- the equipment monitored, i.e. location, type, model, quantity, etc.;
- the duration, frequency, interval, and seasonal or other requirements of measurements;
- personnel, dates and times of measurements.

The more comprehensive BEMS software supports M&V by including four stages of use for the gathered data:

- Awareness: which sets a consumption baseline and can look at power quality;
- Efficiency: which allows the operator to make incremental and proactive behavioural, control and equipment improvements, such as the load aggregation and rate analysis;
- Optimisation: which models production using energy as an economic variable. For this stage, production metrics, regulatory reports and behavioural and climate forecasts would be factored in;
- Aggregation: which will compare consumption information against production output data and other resource-planning-level information.

Models may be explored by using the software to provide evidence for savings to the bottom line based on different plans of action. The energy manager would use this information to build a case for investment in efficiency to senior management.

The rest of this book is concerned with suggestions for how the details of the plan could be tackled, by order of topic, for a wide range of situations.

Energy managers share their experience

Ashley Baxter

What constitutes a good energy manager?

Ashley Baxter won the title of Energy Institute's Energy Manager of the Year in 2006. At the time he was working for Nottingham City Council, England. He is now the director of Hydrostat, a consultancy that helps firms and organisations reduce the environmental and cost implications of their energy and water use.

In his own words, these are the factors which impressed the judges:

Firstly, I shifted almost all the electricity contracts over to 'green' supplies. This made Nottingham City Council one of the largest independent purchasers of renewable energy in Europe, and achieved at negligible cost to the Council.

Secondly, my team and I were responsible for the installation of over 200 half-hourly electricity meters. These not only enabled us to identify waste and unusual consumption, but also to tell everyone about it. The local paper sent the story to the front page – 'Energy Spies To Slash City's Bills'. As far as the *Nottingham Evening Post* was concerned I was Big Brother!

Thirdly, I was responsible for reducing overall Council water consumption by over 25 per cent. This was achieved mainly by introducing sophisticated urinal controls and leak detection equipment into the Council's ageing building stock. The controls were particularly effective in schools where the practice of allowing pipework and fixtures to leak and flush throughout weekends and holidays was brought to an end through some fairly easy technical fixes. The saving on the Council's water bill is worth more than £300,000 a year.

Finally, and most importantly, I encouraged the Council to embed environmental performance indicators into the service planning process. This was the real litmus test of whether the Council was prepared to walk the walk on sustainable development. It wasn't easy. Alongside my constant whinging, there were two other main drivers in 'changing the system', namely a Councillor scrutiny group into sustainable development and also the Local Authority Carbon Management Programme (LACM) coordinated by the Carbon Trust.

Leading LACM for the Council was quite the most challenging project I was involved in. It involved collating a lot of data that should have been easy to find but wasn't, and then using it to persuade managers, architects and engineers that improving environmental performance was both desirable and necessary.

3

Making change happen

The remainder of this book will be about technical and financial issues, but in this chapter we are going to look at an often overlooked issue: the psychological angle. An energy manager has to be many things, and a technical adept is just one of them. Among the other skills they should have under or in their utility belts are: business acumen, IT wizard, psychologist, publicist, and diplomat. All of these skills will need to be deployed at various times in the design, implementation and evaluation of an energy management plan that is to stand any chance of being a sustainable success.

Many professionals acknowledge that how you get everybody on your side – in other words, encouraging behaviour change – is often just as important a factor in success as being technically accomplished. For it to have a lasting effect on the nature of the business, energy efficiency must become a core part of the way in which an organisation sees itself.

Appraising the state of play

There are many challenges faced by leaders of organisations wishing to institute change. Where does resistance to change lie, and what strategies are available to overcome it? Is it necessary for there to be a crisis to motivate change, or is action more effective during periods of stability? How is change best institutionalised? How does an organisation implement sustainability goals while surviving in a potentially conflicting competitive market? How can you tell the difference between 'greenwashing' and genuine action?

To answer these questions, it is useful to use the organisational energy management appraisal matrix in Table 3.1. Such a matrix may be employed as part of a seminar or discussion that is aimed at determining the present level of energy management policy and implementation within an organisation. Each participant would be invited to identify where they think the organisation currently stands. They do this by looking along the columns for each of the levels, beginning with level 0, and identifying in each row, for each topic, how they think the organisation performs. For instance, if there is an energy manager already in place, but they also have responsibility for other matters, then in the row 'Organisation' the organisation is at level 1. If some staff training has been implemented on an ad hoc basis only, then in the 'Planning' row they are on level 3. And so on.

Representatives of the organisation then need to determine how great their ambition is. Do they want to achieve level 4, the top level, or will they be happy

Table 3.1 Organisational energy management appraisal matrix. This may be used to allow an organisation to position its current state in order to help design a new policy and action plan. Most organisations will find they are on level 0 or 1. Progressive organisations are on level 4

Level	0	1	2	3	4
Policy	There is no explicit energy management policy.	An unwritten set of guidelines.	An unadopted energy policy is set by the energy manager or service department manager.	A formal energy policy exists, but there is no active commitment from top management.	An energy policy, action plan and regular reviews exist. Top management is committed as part of an overarching environmental/ sustainability strategy.
Organisation	No energy management or any formal delegation of responsibility for energy consumption.	Energy management is the part-time responsibility of someone with little authority or experience.	An energy manager is imposed, reporting to an ad hoc committee, but line management and authority are unclear.	The energy manager is accountable to an energy committee representing all users, chaired by a member of the managing body.	Energy management is fully integrated into management structure. There was clear delegation of responsibility for energy consumption.
Communication	No contact with users.	Informal contact with engineer and a few users.	Contact with major energy users through the ad hoc committee is chaired by a senior departmental manager.	The energy committee is used as the main channel together with direct contact with major users.	Formal and informal channels of communication are regularly exploited by the energy manager and energy staff at all levels.
Information	No information system. No accounting for energy consumption.	Cost reporting is based on invoice data. An engineer compiles reports for internal use within the technical department.	Monitoring and targeting reports are based on supply meter data. The energy unit has ad hoc involvement in budget setting.	Monitoring and targeting reports exist for individual premises based on sub-metering, but savings are not reported effectively to users.	A comprehensive system sets targets, monitors consumption, identifies faults, quantifies savings and provides budget tracking.
Planning	No promotion of energy efficiency.	Internal contacts used to promote energy efficiency.	Some ad hoc staff awareness training.	There is a programme of staff awareness and	The value of energy efficiency and the performance of

Level	0	1	2	3	4
				regular publicity campaigns.	energy management is publicised both within and outside of the organisation.
Investment	There is no investment in energy efficiency in premises.	Only low-cost measures are taken.	Investment is made using short-term payback criteria only.	The same payback criteria are employed as for all other investments made by the organisation.	There is positive discrimination in favour of green schemes with detailed investment appraisal of all new build and refurbishment opportunities.

Source: Carbon Trust

with reaching level 2, say, in two years, and level 3, two years after that? In order to inform this decision, some information will be required, in particular the current baseline position in terms of energy use, and the costs of taking different forms of action. This is, in effect, a business plan, as outlined in stage 3 of the eight stages of producing an energy management action plan, as outlined below.

The importance of allocating responsibility

Time, money and lack of awareness are the chief barriers to action on energy efficiency. In order to surmount them, it is crucial to ensure that the CEO is accountable for sustainability performance. By elevating responsibility to this level, all parts of an organisation will understand that it is considered to be a core business concern.

For those energy managers unfortunate enough not to be working in such an organisation, a business case needs to be made to management in order to persuade them to sit up, take notice and allocate precious resources to what has until now been a lower priority than whatever is the core business. After all, the cost of energy affects the bottom line, which affects profit. This is covered in Chapter 10. These days, many organisations also have a legal commitment to report on their carbon emissions, the biggest ones being subject to penalties should they not reach certain targets. Furthermore, commercial advantage may be obtained through having a positive public image in terms of friendliness towards the environment, energy saving and sustainability. That is why it is vital that any successes are publicised properly to all stakeholders.

The eight stages of an energy management action plan

Figure 3.1 shows the eight stages of an energy management action plan. The most important component of this is the people who are being engaged by it, and the most important skill required of the energy manager is being able to motivate them. Their personal qualities will be tested to the limit, and if they don't feel confident they should engage the support of someone who can help them in this aspect of their work, or seek some kind of training. Let us look at each stage in more detail.

Figure 3.1 The eight stages of an energy management action plan.

Source: Author

1 Making a commitment

The first stage is for the organisation to make an explicit commitment to energy efficiency and the reduction of carbon emissions. This usually involves writing an energy policy: a short, written statement of senior management's commitment to managing energy and its environmental impacts. Often it forms part of a wider corporate social responsibility (CSR) policy. For large organisations this document is usually no more than two pages long; for smaller ones a few paragraphs may be sufficient. It requires an active decision and commitment from senior management level, with the willingness to allocate a sensible amount of resources.

2 Establishing the baseline

This is what we have seen in Chapter 1: the production of an initial energy audit to show where energy is being used throughout the organisation at present and, possibly, the M&V plan. This should be quantified in terms also of financial cost to the organisation, and carbon emissions. It should be clear where most of the energy is going and why. It should be benchmarked to the performance of similar organisations.

3 Constructing a business case

A financial case will present options for action that may then be evaluated and decided upon by senior management so that they may allocate resources for a campaign. It will ideally contain several contrasting scenarios for energy savings that can be secured over different time periods, dependent on the upfront amount of investment of capital and employee time. Therefore it will show the level of return on investment in each case. To construct these scenarios it will be necessary to know what future plans the organisation has for expansion or contraction. Projected energy use will then be a function of these plans. It will be necessary to have an idea of the future cost of energy. While this is always hypothetical, governments have figures that they use themselves in their forward planning and, for want of any other source of information, this is what may be used.

One of these scenarios will be 'business as usual', which is to carry on without doing anything different. Perhaps three other scenarios could be imagined and quantified: 'deep green', 'medium green' and 'light green' options. Included in these might be some degree of carbon offsetting and some degree of sourcing energy from renewable sources. However, it must always be borne in mind that it is far cheaper in general to reduce energy use than to invest in new renewable energy generation plant.

Each scenario will contain the following information:

- the baseline level of energy cost gleaned from the energy audit or assessment;
- a figure for the level of financial investment to be made;
- a figure for the level of energy savings that may be achieved for this amount of investment, given in megawatt-hours or Btus, and the associated carbon emission reduction figures, together with cost savings that will be achieved;
- the period over which the savings would be made;
- the chief actions which can bring about these savings.

The organisation will consider these options, they will inevitably be modified, and finally the favoured option will be chosen as a basis for creating an action plan.

What level of resources should be allocated?

This is usually decided with reference to the annual energy spend of the organisation. As a general rule of thumb, spending of each £1 million (US$1.4 million) per year will justify the employment of one full-time-equivalent post, since it will easily pay for itself in savings. For example, if £4 million (US$5.6 million) is spent a year, then four full-time-equivalent positions are justified. This figure includes the resources to be made available to them, but if there is to be major capital expenditure (e.g. new plant or a significant refurbishment), then this will obviously be additional expenditure.

---

4 Create an action plan

An action plan for an energy strategy, essentially the M&V or energy management plan, sets out how energy will be managed in an organisation over a period of time. In order to secure buy-in of as much of the organisational staff as possible, everyone should be allowed to contribute suggestions that may be incorporated within the plan. A top-down, management-imposed plan will have little chance of succeeding.

The plan establishes the management framework, the positions of responsibility, and allocates tasks to individuals or posts within that structure. Each task has a measurable outcome at particular staging posts. This allows for the monitoring of the plan's implementation. It will make reference to:

- roles and responsibilities, including where the final responsibility lies;
- whether a formal management system is used, such as ISO 14001, BS EN 16001, or ISO 50001;
- resources available;
- energy procurement;
- compliance with energy and climate change regulations;
- investments required;
- implications for procurement strategy;
- implications for staff behaviour and training/awareness raising;
- the strategy for metering, monitoring and targeting (MM&T);
- the strategy for review, feedback, employee motivation and publicity.

An annual review may be sufficient in some cases, but more frequent reviews, especially at the early stage, are desirable to maintain momentum and ensure the efficient allocation of resources.

5 Appoint energy champions

There is nothing like assigning responsibility to enthusiastic individuals for achieving results quickly. No one knows better than the people on the ground where waste occurs. This is particularly appropriate where there are different buildings, facilities or sites, or it may be that different departments have their own energy champion. Volunteers should be encouraged to come forward. In an ideal world they will have been involved in the creation of the action plan. If not, they will be using it to come up with their own suggestions as to how its targets can be achieved in their area.

Again, their responsibilities should be clearly set out, with targets and timescales. For instance, they can make sure that windows are closed when they should be, lights are switched off, equipment switched off at night and weekends, the thermostat appropriately set, taps not left running, maintenance carried out on time on the building fabric, etc. They might also keep an eye on sub-metering in their section.

Results may be collected weekly or monthly and shared among all other energy champions. A section of the corporate intranet may be allocated for them to exchange advice, best practice, results and resources. An element of competitiveness may be encouraged, so that they can compare their performance with each other. Ideally, there should be rewards for positive results. These don't have to be financial: it could be a simple as personal recognition within the company, or being given a cake (as proven to work by More Associates, in their Carbon Culture initiatives (see DECC case study on p. 49).

Case study: Sainsbury's

Supermarket chain Sainsbury's saved up to 5 per cent on energy consumption by giving one staff member in each store the responsibility of driving energy efficiency measures for that store. Their brief was to understand how their store uses energy, and identify where savings could be made. Part of their responsibility was reporting progress back to their colleagues at each store to raise everyone's awareness of what they could do. This created a feeling of ownership at store level, thus ensuring that the measures would not only be implemented but maintained. Everyone was encouraged to come forward and make suggestions.

These suggestions included easier ways to control lighting at the rear of the stores, a process for giving staff daily updates on energy use, and using sensors and timers. As a result of the trials, energy management is now part of the job specifications for commercial managers at Sainsbury's stores together with additional training.

6 Implement the plan

In medium to large organisations, the plan is carried out by an energy management team under the responsibility of the energy manager. Their responsibilities include the ongoing monitoring and reporting on energy use, its cost and related carbon emissions through the use of M&V tools. They will benchmark performance, identify exceptions and instigate corrective actions.

The plan includes a programme of regular communication with staff to encourage and motivate everyone to be energy aware and contribute to meeting the targets. Support and advice will be made available. Sometimes staff such as engineers or building managers may declare that they have conflicting instructions from line managers or their job description. This is an indication that the plan has not been completely integrated within the organisational structure. The energy manager must therefore go up the chain of command to try to resolve this type of dilemma and achieve a compromise.

On an ongoing basis it will be necessary to keep up to date with new regulatory requirements, technical developments and sources of funding for energy efficiency investment, at the same time as identifying and acting on opportunities to reduce energy consumption and use lower carbon sources of energy.

The energy manager should have an influence on procurement decisions so that there is a sustainable procurement policy. Whenever maintenance, replacement of technology, equipment purchasing, building refurbishments and new builds are to occur, the energy manager should be informed sufficiently far in advance that they can have an input into the specifications and decision-making process, especially so that lifetime energy costs are taken into account. This should have been written into the original commitment.

All communications between the energy manager and those on the ground, such as a building manager, need to be part of specified processes. Responsibilities and tasks should be formally allocated. If they are in any way informal, they are unlikely to be acted upon.

In many cases, interventions work best when choice is taken away from individuals, who, no matter how well motivated, will often simply forget. Thus installing energy-saving software on PCs, automatic temperature and humidity controls, variable speed motors, and motion- or infrared-sensors for lighting controls will quickly pay for themselves.

But there will inevitably be many areas where human behaviour is a key factor. This type of intervention should be staged and not all happen at once. This is because staff will inevitably quickly forget, no longer notice posters, lose information and delete e-mails. A regular newsletter, alerts, staged actions, etc., are all ways of continuing to refresh the campaign. For example, in the winter focus on heating, and in the summer cooling could be the topic. At other times lighting, PCs, photocopiers and so on could be topics. If interventions include changes to the building fabric (e.g. replacement of windows, lighting, draught-proofing, insulation), this may be used as an opportunity to raise among staff their awareness of the importance of energy efficiency and saving energy as well as improving employee comfort.

It is important to consider what motivates people. There has been much work on energy saving and psychology. For example, if staff are told that savings made would be reinvested in improved facilities or donated to a nominated charity, this can be a powerful incentive. They may also be motivated by actions which improve the quality of their working environment. The provision of a suggestions box is often a good idea. It is important that communications use the right kind of language and hit the right tone. If not, they can be terribly off-putting. It may be a good idea to trial or test posters or messages with a sample group before broadcasting it widely. The use of appropriate metaphors and imagery can be particularly appealing. For example, using imagery and metaphors that are to do with football, baseball, popular TV programmes, or other ideas that are already accepted as part of the workforce's self-identity will encourage take-up.

Some organisations swear by using e-mail and the intranet to disseminate messages. Others use meetings and staff training, but staff may be too busy to attend these events. Posters and newsletters are good for some, but may be ignored by others. Some messages will only be read if they are included in a payslip. In other cases these channels may completely miss out many individuals. Some people respond best to being directly addressed and told information or shown what to do face-to-face. It is therefore important to select the right channels and, perhaps, to have multiple channels to ensure that the messages reach everyone.

Some people respond to the challenge of a competition, say, between buildings, departments or sites, on who can make the greatest energy savings. The results can be published and, if possible, a prize awarded to the winner. Some organisations run an Energy Week every year to maintain awareness and progress. Competitions may also be reinvented with different targets and content.

Case study: UK Department for Energy and Climate Change Offices

This government department commissioned external consultants to design both an engaging way of displaying real-time energy use, and a suite of novel social-media-style tools to engage users in competitions, points and prize winning and information exchanging on reducing the department's energy use. Staff were encouraged to tell their stories about their own energy use. It was found that awareness of the actions of others is a strong driver in getting more people involved in new actions and helps communicate everyone's progress to each other. Real-time energy displays sparked conversations about how energy was being used and introduced a sense of ownership over the peaks and troughs in the graphs that represent energy use. It also enabled the sustainability and estates team to optimise the building's heating and cooling systems. Ten per cent of gas savings were achieved in under a fortnight.

Purely using this technique, the pilot found a high level of uptake for this kind of engagement process – 40 per cent of staff – and secured 20 per cent daytime gas savings for heating, although it is unknown for how long this was sustained. Different users were found to be attracted or motivated by different tools. As this was a research programme and its effects were being monitored, it was found that certain tools pulled in additional groups of users at different points during the campaign. In addition, the order in which this was done made a difference, which helped reduce costs and maximise overall engagement of staff.

The technique of adding further options and further topics and incentives over successive weeks worked to continually refresh the programme and give an uplift in participation numbers. This type of campaign will only work in offices where staff already use an intranet or the internet as a habitual part of their daily activities. In such a case carbon and energy savings may be achieved at a relatively low cost compared to other types of campaign. The campaign went on to be rolled out in many other government departments, including Number 10 Downing Street. www.carbonculture.net/research/decc.

7 Monitor and evaluate progress

It is vital that periodic checks are compared against baselines and predictions for energy savings that were made in the models, or against the targets that were set. While some margin of error is to be expected, deviations should be examined both to take the chance to improve the modelling, and to improve the interventions and actions taken. At regular periods and at the end of a campaign, results should be distributed among all staff, management and stakeholders. This is particularly true at the end of the campaign. If staff members are aiming for specific targets, let them know how close they are, using easy-to-understand methods. A display in the foyer is an effective way to keep everyone informed and up to date.

Included in the messages are lessons that have been learned and new initiatives to be taken as a result. Every attempt should be made not only to engage everyone, to give them a sense of ownership in the project, but not to apportion blame or make anyone feel guilty for not participating or making erroneous choices. This would be a powerful disincentive to participation, and encourage resentment.

Above all, the aim should be to generate a sense of collective pride in individual and group successes in reducing costs and carbon emissions. Staff should be encouraged to share their experiences, both their successes and issues that they have found difficult. If they have encountered a problem, invite others to come forward with solutions. It is not necessary for the energy manager to appear to be the one with all the answers. For ownership of the campaign to be shared and to maximise participation, the viewpoints and contributions of everyone should be valued. Conversations should be encouraged so that a culture of positive change is brought about. There are times when it is best for the energy manager to step back and let others try things out, do their own research, and learn from mistakes. In this way, not only may novel solutions be discovered, but people feel that the knowledge gained is theirs. They feel a sense of achievement.

The energy manager should at all times be aware of how the timing and tone of the campaign fits in with the other things that are happening within the workforce. Staff morale may be boosted by feeling empowered by the campaign, or, if they feel discouraged or disinterested it can have a negative effect on overall morale. Moreover, if morale is low for some external reason, this may not be the best time to get people on board with a new stage of the campaign. Timing is everything.

If all of this sounds naive or overly optimistic, consider that it is not necessarily as difficult as it may seem. A very simple campaign could have the following structure: it lasts 24 weeks, and every four weeks a new aspect of the campaign is launched, focusing, say, on heating or cooling controls, lighting controls, windows, water use, IT use and transport. A dedicated noticeboard is set up in the foyer or canteen.

Say there are five teams competing, whether they are in different buildings, on different floors or in different departments. Every week, on a Monday morning, they receive the results of the previous week's energy use and they can all see how each team has done. This may come in the form of a single sheet of paper or a folded newsletter, either printed or by e-mail. At the same time, if any questions have been asked by any member of staff about what they can do, these are included in the newsletter, but no answers given. A channel is set up for responses, whether by e-mail, by post-it notes or a suggestions box by the noticeboard. The following week, any answers to these questions that have been forthcoming from other staff members are also included in the newsletter.

Every four weeks, the next topic is introduced at a 45-minute meeting, with coffee and biscuits, on the Monday morning, as well as through the distribution of newsletters. If a meeting cannot be arranged, the same information may be posted on the dedicated noticeboard and circulated by e-mail. What should happen is that monitored energy use is reduced step by step in order of the interventions that might be made as a result of actions on each topic.

8 Recognise achievements

In the above example, at the very end, prizes are awarded and recognition given to the staff members or team that have achieved the most. At the same time everyone is recognised by having put in an effort, since some people will have found that steps they wanted to take were impossible for reasons outside of their control. In fact, these discoveries will highlight areas of action for future campaigns, but since they may involve substantial investment in new equipment or insulation, that may require new budget. Publicising and recognising the actions and successes which staff make will encourage them to make further suggestions.

Overall, the campaign would need to be written up and the results disseminated both at board level for inclusion in annual reports, and to the general public and other stakeholders, perhaps by the press and publicity office. Not only will it reflect well on the organisation that it is taking these steps, but other organisations will be able to learn from what has been done.

Energy professionals share their experience

Lisa Gingell, Director

Lisa is the only female member of the Energy Services & Technology Association (ESTA) Council, as well as director of t-mac Technologies Ltd, which specialises in internet-enabled remote data logging and control devices with GPRS connection for wireless installation for Building Energy Management Systems.

What does your job involve?

A large part of my role is understanding the current market and identifying where changes can be made. We are focused on the 'demand' management of energy use through metering, monitoring and BEMS/controls, and subsequently legislation, regulations and market changes are of paramount importance.

What qualifications did you gain?

I hold a BA Hons degree in social sciences. This has proved very useful in my role, particularly the psychology aspect. Social sciences is all about understanding behaviours and the way people and the marketplace change; understanding how you can influence and bring about behavioural change ultimately paves the way for new thinking and new requirements in energy management.

What does your day-to-day work look like?

Most days are split between running the business, managing key client accounts, market research and developing product road-maps. I am also in charge of marketing and communications, which involves developing and implementing PR and advertising strategies, planning for and attending exhibitions, hosting seminars, and delivering workshops. I am also very active in trade associations. I was recently elected to the Energy Services & Technology Association (ESTA) Council. This is a great honour, especially as I am the first female council member in its 30-year history. ESTA is the UK's leading energy management industry association and I bring my market knowledge and marketing expertise to the role. I'm looking forward to contributing to the growth of the organisation. We also recently launched a new campaign, bringing social media to the energy industry.

What do you love about your job?

Constantly coming up with new technologies and targeted marketing strategies. We are just about to launch a new energy iPhone app – a first for the energy industry – putting energy in the palm of your hands!

What do you think are the biggest challenges?

Legislation tends to make the industry focus on the necessities and ensure boxes are ticked. Unfortunately this often does not translate into monetary savings. In an economically challenged marketplace, focus for energy management needs to be on return on investment and this can only come from monitoring *and* controlling how a building operates its energy – from equipment to building occupiers themselves.

What's the best way to engage people on energy efficiency?

You have to understand what really makes them tick. Building occupiers come from all walks of life. But people respond to visual data that is appealing, interesting and engaging. Focus on the message and understand the audience. Then create a solution that suits them; not a one size fits all. We offer a variety of solutions from analytic tools for the advanced user through to the iPhone apps and custom energy-showcasing dashboards for the building occupiers/employees to engage the different audiences. They are all using the same data and running on the same software platform but their messages and medium are very different.

What is the first thing that you would recommend to save energy and carbon emissions?

Measure and monitor, analyse and quantify, then add *controls* – be it automatic controls through BEMS or manual controls by engaging building occupiers to get involved; it is a simple formula for success.

What is your favourite tool of your trade?

It would have to be our system – it does everything we think you need in this modern day and age for effective energy management. It provides real-time metering and monitoring, connects to equipment in the building and controls use, to ensure the building, and equipment in it, are operating to an optimum level with energy efficiency in mind. The online software covers the analytics as well as the showcasing of data and developing engagement platforms for the building occupiers and energy managers.

4
Airtightness and insulation

The energy consumed in heating and cooling a building is a function of its heat gains, the quality of the building fabric and the behaviour of the building's occupants. The previous chapter discussed the latter aspect; we shall now turn our attention to the first two.

Heat arises within buildings through the following ways, which are called 'passive' heat gains:

External loads

- orientation;
- window area;
- glazing properties;
- shading system;
- insulation level (U-value or R-value: See Appendix for explanation);
- ventilation level.

Internal loads

- occupants;
- lighting;
- equipment.

Heat leaves or enters buildings in the following ways:

- uncontrolled draughts, via gaps in the building envelope;
- controlled ventilation;
- conduction through the surrounding material of the building, its roof, walls, floor, doors and windows, otherwise known as the building envelope.

Depending on the exterior temperature, this may either result in unwanted heating or cooling within the building. All of these factors need to be measured and controlled if the aim is to reduce the total amount of artificial energy input required.

Figure 4.1 Factors
affecting internal
temperature.

Source: Author

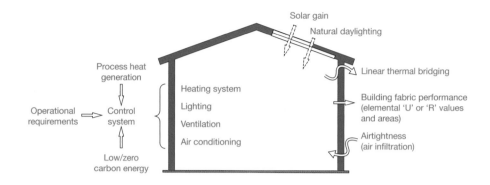

Low or zero carbon buildings

Nowadays the trend is increasingly towards low carbon or zero carbon buildings. The American Council for an Energy-Efficient Economy projects that in the medium-term future no large commercial buildings will need any heating.[1] The strategy advocated in building regulations to achieve this is to minimise the energy required to cool or heat the building through superinsulation and airtightness, at the same time maximising the solar gain for heating, or minimising it in a hot climate. Doing the former keeps the sun's free heat within the building.

Following this, should that strategy alone not achieve a zero energy requirement, any further energy requirements would be satisfied from heat reclamation or renewable energy, either on- or off-site. This is covered in Chapter 9.

The stack effect

The 'stack effect' is used to help circulate heat in a building and moderate its internal temperature and climate. It uses the principle that warm air rises by convection to exit at the top through open windows, vents, chimneys or leaks. This causes reduced pressure lower in the building, drawing in colder air through any openings. Good passive solar design takes advantage of this principle. In large or high-rise buildings with a well-sealed envelope, pressure differences can become acute. Steps should therefore be taken to mitigate this, with the help of mechanical ventilation, partitions, floors, and fire doors to prevent the spread of fire.

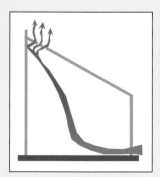

Figure 4.2 Illustration of the stack effect. Warm air rises, and warm air leaving the top of the building will draw in cold air at the bottom where it can.

Source: Author

In a superinsulated, passive solar building, ventilation is controlled in a way that permits the occupiers to regulate the air's humidity, temperature and cleanliness, otherwise known as the internal climate. Ventilation may be passive, using the stack effect, if available, or active, using pumps and ducting. In a low or zero carbon building, the pumps include a means of recovering the heat from the air leaving the building and passing it to the incoming air. This is known as ventilation with heat recovery and is covered in Chapter 6.

There is a limit to what may be achieved with existing buildings in terms of insulation and airtightness. Budget restrictions will force an additional limit. For this reason, every opportunity should be taken to work with employees to be alert to any occasions when refurbishment is already scheduled, as this will be an opportunity to introduce energy-efficient measures such as extra insulation most cost-effectively.

A major refurbishment, or weatherisation, of an entire building is especially an ideal opportunity to reduce heating demand almost completely through super-insulation, because the additional cost of doing so will be much less. According to the US Federal Energy Management Administration (FEMA), 'the effective life of an office building is 20 to 30 years, after which major renovation and updating is normally necessary'. This additional cost must be justified on a building lifetime cost-saving basis. In other words, the business case is to calculate the lifetime saving on energy bills gained from these additional measures and use them to offset the additional cost of the work. Expressed as an interest rate return on investment, this is usually a powerful argument (see Chapter 10 on how to achieve this).

In the absence of a refurbishment opportunity, the first step to reduce heating or cooling energy demand is to remove unwanted airflow.

Draughtproofing

The quickest and cheapest ways to retain heat in a cold climate or during cold weather, or, in the context of a hot climate, keep out heat, are achieved by reducing unwanted airflow in and out. A check should be made of each room, and the exterior where safe. All unwanted openings must be sealed, such as:

- around doors and windows;
- where services enter or leave the building;
- gaps in flooring;
- around the edges of walls at floor and ceiling level;
- airbricks;
- trickle vents;
- construction joints between materials;
- at the eaves.

Dampers should be fitted to any chimneys flues, and intermittently running extractor fans should have well-fitting, self-closing covers which securely close when not in use.

Special attention should be paid to fittings on doors and windows. Badly fitting doors and windows are a major source of draughts. Small gaps around the frames

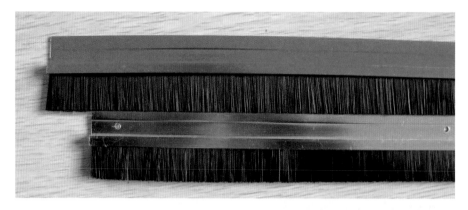

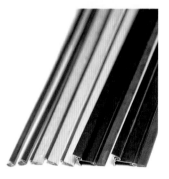

Figure 4.3 A variety of common draught seals: brush seal, wiper seal, compression seal, service seal.

Source: Energy Saving Trust (first three); Chris Twinn (service seal)

may be filled with gun-applied sealants and fillers. Draught-stripping around the openings themselves is inexpensive, simple to install and greatly improves comfort as well as reducing fuel bills. Many different types of seals are available, including compression seals, low-friction or wiper seals and seals for larger gaps around frames.

Windows and doors

Some doors or windows may need to be replaced. Existing single-pane windows and windows with metal frames should definitely be replaced. If this is not possible, secondary glazing may be fitted. Secondary glazing fits inside the reveal, and may be removed for cleaning purposes. The edges should be compression sealed. Replacement doors and windows should have insulated cores (i.e. with insulation between the two outer surfaces to prevent thermal bridging), and preferably be constructed from timber for environmental reasons. These measures typically pay for their cost in the value of the energy saved in three to four years.

The glazing should have a low-emission (low-E) coating to allow light to pass through and reflect infrared wavelengths back in. When procuring glazing the amount of solar energy which the window lets in can be controlled by specifying the type of coating on the inside of the first pane of glass.

Controlling passive gain

Windows are essential to provide free lighting and heating, but this must be controlled so that occupants do not experience glare and the building does not overheat. Solar gain is retained whenever the sunlight falls on to a surface that

Left: Figure 4.4a Three levels of seal and a top-performing handle/catch system guarantee no draughts on this Passivhaus-certified door.
Below: Figure 4.4b Insulation within the frame and between the panes of a triple-glazed window or similar door removes thermal bridging and contributes to the Passivhaus level of performance.

Source: Manufacturer

will absorb and retain the heat in the thermal mass behind it. Carpets or curtains covering this absorption surface will prevent it from doing so.

Thermal mass is the product of the specific heat capacity of the material and its total mass and conductivity. The more dense a material, like stone or concrete, the more heat it will hold. Traditionally, in hot countries, adobe or rammed earth performs the function of preventing overheating. The wall acts predominantly to retard heat transfer from the exterior to the interior during the day. Its degree of success is highly dependent on marked diurnal temperature variations. Thermally dense materials take longer to heat up and to cool down afterwards. In cooler climates, having warmed up during the day they release their heat into the internal space overnight, moderating extremes of temperature.

Hotter climates

Figure 4.5 Exterior shutters used in a hot climate to keep out unwanted heat in the daytime. They are opened in the evening if required.

Source: Author

In the US region of Sacramento, California, a west-facing window of just $5m^2$ ($55ft^2$) will add as much as 16kWh (55,000Btu) to a building on a summer's day. Compensating for this with an air conditioner would require almost 2kW per hour. This could be saved by the use of an exterior shade for a fraction of the cost. The following features for exterior walls and roofs will help prevent the building from overheating:

- Whitewashing or painting exterior surfaces a pale colour;
- External shading calculated to keep windows in the shade during the summer months;
- Making the building wider on the east–west axis than on the north–south axis to minimise the solar gain that enters beneath the overhangs in the morning and afternoon;
- Having no skylights and roof windows, to keep out the midday sun;
- Keeping shutters and curtains closed during the day and open at night;
- External insulation.

External shading may be provided by trees, which have an additional cooling effect. Deciduous trees shed their leaves in the winter, allowing the sun's heat into the building. Fixed architectural elements for shading include, besides overhangs: pagodas, vertical fins, balconies and false roofs. External shades are about 35 per cent more effective than internal ones. Overhangs are sized relative to the latitude, location and window size. Internally, light shelves and louvres may be used. Adjustable elements include: awnings, shutters, blinds, rollers and curtains.

Figure 4.6 This pergola provides shading to prevent overheating. The French windows are Passivhaus certified.

Source: Author

Figure 4.7 External shading devices added to the outside of a highly glazed south-facing office block to prevent overheating in the summer months but to maximise solar gain at other times of the year. Welsh government buildings in Aberystwyth.

Source: Author

Figure 4.8 The addition of glazing positioned away from the building exterior allows the sun's heat to warm up the thermal mass of the building's wall, which is then communicated inside to moderate temperature fluctuations and supplement the heating. WISE educational building at the Centre for Alternative Technology, Wales.

Source: Author

Figure 4.9 The addition of an overhang over a window can prevent glare and heat gain in the summer, but permit low-angled sun to provide light and warmth in the winter.

Source: Author

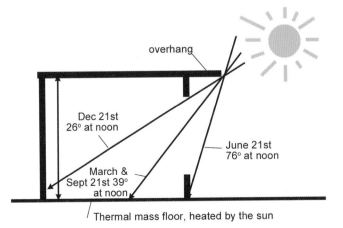

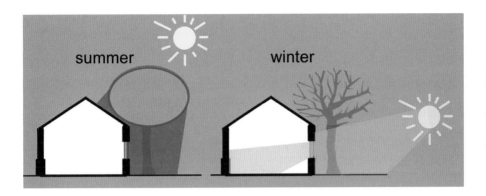

Figure 4.10
Deciduous trees may be planted outside to prevent overheating or glare in summer. In winter they lose their leaves and low-angled sun may provide extra warmth.

Source: Author

Cooler climates

In cooler climates, similar principles apply but they are applied differently. Here, Equator-facing windows on the east–west axis will permit a large amount of glazing to be used in relation to the building's volume. Skylights and rooflights can be beneficial as long as precautions are taken to avoid overheating and to avoid heat loss at night and during cold periods. The sun's light should be permitted to fall on to a thermally massive surface, such as a wall.

Building design can allow this direct solar heat to be conducted, radiated or converted to other parts of the building, a topic addressed in Chapter 5.

Automatically adjusting motorised window-shading-and -insulation devices may be installed that are controlled by sensors that monitor the temperature, sunlight, time of day and room occupancy. These devices can reduce by over half the energy requirement for cooling.

Superinsulation and airtightness

Figure 4.11 Poor building practice like this leads to air gaps and thermal bridges that would later be covered up and invisible to a visual inspection. External insulation is the simplest way to rectify such problems.

Source: Stamford Brook report on airtightness, courtesy Malcolm Bell, Leeds Metropolitan University, reproduced in EST (2008)

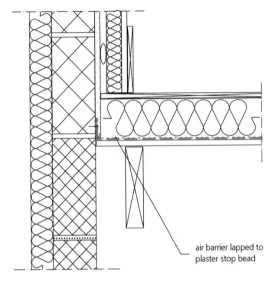

Figure 4.12 Detail showing the position of an airtightness barrier (blue dotted line) beneath the insulation under a suspended floor, lapping up the side to meet plaster/plastboard. On the inside of the plaster is more insulation. Insulation is also on the outside of the building. Together this makes the building thermally insulated and airtight. Diagrams like this are free from the enhanced construction details (ECDs) section of the Energy Saving Trust website.

Source: EST

air barrier lapped to plaster stop bead

Simple draught-proofing techniques provide quick gains, but introducing airtightness and superinsulation, whether in a cold or hot part of the world, will maximise the potential for control over the internal climate in order to minimise the energy cost. At the same time, an airtightness barrier can be introduced around the building envelope. This is a continuous layer that surrounds the building. The idea is to prevent unwanted draughts through often invisible or inaccessible holes in the building fabric; for example, underneath floors or where mortar has not been applied consistently in between blockwork, or where joists enter the exterior walls.

What can the airtightness barrier consist of?

Vapour permeable (and hygroscopic) materials

Vapour permeable membranes, intelligent membranes, lime, concrete, timber, hemcrete, bricks, stone, plaster, mineral (rock and slag) wool (Rockwool).

Non-permeable materials

PVA and vinyl paint, when used as a coating on plaster, metal, glass, PVC, plastic, plastic foams, XPS, EPS, phenolic foam insulants.

If problems with damp and condensation are present or anticipated, moisture must be able to pass through this layer; this is not possible with non-permeable materials. Instead, the materials used should be hygroscopic and vapour permeable. Of course, not all of the building envelope can be vapour permeable, the obvious example being glazing. If most of the exterior building skin is impermeable, as in the case of buildings clad in sheet metal or insulated with polystyrene or foam-based insulants, condensation problems may be resolved with the use of a dehumidifier in the HVAC system, but this adds additional energy load.

Whatever combination of materials comprises the building envelope, there should be no breaks in it; for example, tears in membranes, gaps in masonry, around windows and doors, or between insulation, or nail puncturings. Doors and windows must be properly sealed when shut. Professional long-lasting tape must be used to seal joins in membranes. The airtightness barrier must be able to withstand pressures created by wind, stack effect or ventilation systems. For large structures, every floor or occupancy unit should be treated as its own airtight area.

Regardless of the best efforts, a building is never completely airtight, nor should it be. Building regulations specify the number of air changes per hour permitted for safety. The introduction of fresh air is managed with controlled ventilation. The level of airtightness is tested by professional pressure-testing devices, called blower doors. The common metric is the number of air changes per hour at a specified building pressure, again, typically 50Pa (ACH50). The standard of airtightness for Passivhaus certification (see p. 73 for more information) is that a new building must not leak air at a rate greater than 0.6 times its volume per hour, at the standard pressure for blower door tests that are used to measure airtightness in buildings (which is 50Pa (N/m^2)). This standard is around four times better than most building regulations. A refurbished building could achieve the EnerPHit Standard, which is designed to make such buildings close to Passivhaus but reduce costs. With this, the criteria have been relaxed a little so that the airtightness level is one building volume air change per hour.

Installing the airtightness barrier is easier if the whole building is being renovated, but if this is not possible it is a good idea to plan for the future on the assumption that eventually the wrapping will be continuous. The devil is always in the detail, and the most problematic details are where the wall or roof is penetrated by windows, skylights, joists, doors and service entries. The simpler the geometry of the building, the fewer junctions there will be, so the less risk of leaks.

Insulation

Commercial and industrial premises come in a huge variety of sizes and shapes. They are also constructed in many different materials, from traditional brick and stone, to modern steel warehouses. Retrofitting insulation on to these requires different strategies for each building type.

The highest possible standard of insulation should be aimed for, but there is a law of diminishing returns. The final standard is determined by budget, the space available and choice of material. It is vital to avoid any gaps in insulation. A detailed discussion of insulation and installation strategies is in the companion

Thermal bridging

Thermal bridging occurs when a relatively conductive material passes through the building envelope from the interior to the exterior. This may be a single-paned window or a concrete lintel in a windowsill, any fixings, window or door frames, joists, services and wall ties in cavities. Sometimes the floor slab is extended through the building envelope. As the standard of airtightness and insulation increases in a building, thermal bridging becomes increasingly significant as a factor in heat transfer. To 'break' a thermal bridge, insulation is added on the inside or outside, or even in between, as seen in the door and window designs near the beginning of this chapter (figures 4.4 and 4.5).

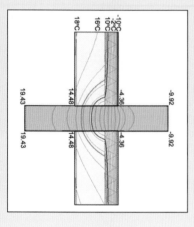

Figure 4.13 A concrete beam passing through from the inside (left, at 18°C/64°F) to the outside (right, at –10°C/+14°F), first through insulation, and then through timber. The concrete beam acts as a thermal bridge passing both heat from the inside outside, and cold from the outside inside. The temperature gradient is modelled in software.

Source: Wikimedia Commons. Author: Bauigel

volume, *The Earthscan Expert Guide to Sustainable Home Refurbishment*. What follows is a brief summary.

Choice of insulation material

EPS, XPS or phenolic foam are often used in the building industry. However, these materials are the product of the fossil fuel industry, and so have caused greenhouse gas emissions during their manufacture, and they are not vapour permeable. There are three recommended, economic types of insulating material for a low carbon building: mineral wool (e.g. Rockwool), wood fibre and recycled cellulose (e.g. Warmcel). The first two, like foam insulants, come as boards and batts. The latter is loose fill and only proven to be suitable for horizontal spaces when contained and uncompressed, in which case it is easy to install, cheap and highly effective.

Mineral wool (k-value: 0.033–0.40 W/mK) has less environmental impact than EPS or phenolic foam board. The latter has a lower k-value ((0.020–0.025 W/mK), so would require less depth to achieve the same level of insulation, but this is its only advantage. Recycled cellulose is made from shredded newsprint (k-value: 0.038–0.040W/mK), and so locks up atmospheric carbon in the building.

Wood fibre board (k-value: 0.80 W/mK), being made from trees, also locks up carbon in the fabric of the building. It may or may not be a disadvantage that twice the depth than of EPS, XPS or phenolic foam is needed to achieve the same level of insulation, depending upon space considerations. Interlocking panels are available specifically for exterior cladding, which eliminates thermal bridging. When rendered, they are weather resistant and breathable. At the end of their life they are compostable.

External or internal?

A decision needs to be taken, first, on whether external or internal insulation should be applied. In general, internal wall insulation may be easier to accomplish, especially one wall at a time, and is ideal to carry out when decorating. It is also cheaper than external insulation. On the other hand, there will be a loss of internal space, it is disruptive to residents, and it is necessary to move electrical sockets and light switches, skirting boards and so on. But, in addition, the thermal heat storage value of the wall is lost, so heat will leave the space quickly along with the air. In other words, it might heat up quicker, but it will also cool down quicker.

External insulation has many advantages. It improves weather protection and noise insulation as well as retaining the thermal mass of the walls inside. Depending on thickness, any value of insulation may be achieved and there is no

Figure 4.14a and 4.14b Tongue-and-groove wax-impregnated wood fibre board cladding over wood fibre batts. The cladding is rendered with lime plaster to make the walls breathable, which reduces the risk of condensation inside.

Source: Author

inconvenience to occupants during installation. Any internal or interstitial condensation and damp is banished, and any gaps and cracks in the wall or poor rendering are covered up. In other words, it is easier to ensure airtightness. Detailing around windows and doors is more easily managed, and it becomes possible to insulate a whole building at once, with consequently lower overheads in spending on scaffolding, etc.

External insulation may need planning permission, and downpipes and other projections or service entry points must be dealt with. Choosing external insulation for a thermally massive structure also implies that the best choice of heating is a constant and low-level one, such as underfloor heating. External insulation involves applying an insulating layer and a decorative weatherproof finish to the outside wall. The aim is to reach U-values below $0.3\text{W/m}^2\text{K}$ or half of this for the Passivhaus standard (see p. 73).

External systems

Three generic types of external insulation are available:

1   Wet render systems
2   Dry cladding systems
3   Bespoke systems.

The first two are often proprietary products for specific situations. Bespoke systems are designed for particular projects and combine elements of proprietary systems, often incorporating dry cladding.

Wet render systems eliminate the need for extensive repointing on brickwork buildings. The better the quality and thickness of the render, the higher performance they are. Traditional render and polymer-modified cementitious render are used for most buildings. Polymer helps to make the render more workable on-site. However, it can suffer badly from inclement weather. Reduced weight can be an advantage in high buildings. These coatings do not need movement joints unless the building substrate has them.

Dry cladding systems are useful where fixings have to be restricted to particular areas and are mostly used on high buildings. The cladding is fixed to the framework. There is a ventilated cavity. The insulation panels are fixed to the substrate with a mechanical or adhesive fixing, or partially retained by the framework. The framework is adjustable, meaning that it can be constructed over an uneven substrate and the finish will be in a true plane. These systems can span substrate areas where fixings cannot be anchored.

Modern façade cladding systems for large and high-rise buildings are often made off-site and installed on brackets fixed to the wall. This maximises the possibility of airtightness. A steel frame holds the insulation, and the coloured render is added either before or after fixing. They can be ordered to specified levels of noise and thermal insulation.

Figure 4.15a, b and c External insulation on terrace apartments in Frankfurt, Germany, with thermographic photographs taken before and after modifications. Blue indicates that there is no heat escaping. Thermographic photography is an excellent a way of revealing heat loss. The design was calculated using PHPP software, which is necessary when designing Passivhaus-standard buildings. It achieved an annual heat energy demand of 17kWh per metre squared area. The strategy was 260mm exterior insulation, triple-glazing, central ventilation heat recovery, reduction of thermal bridges and 7.5m² (80ft²) of solar collectors. A pressure test recorded an air change rate of 0.461/hr.

Source: International Energy Agency

Figure 4.16 External insulation being installed on an old brick industrial building. The insulation is rockwool, made of reclaimed stone wool waste and residue from other industries.

Source: Author

Choice of render

In terms of renders and their environmental impact, hydraulic lime renders are preferred, since they provide a lasting, breathable surface. They consist of around a 12mm coating of hydraulic lime containing glass fibre mesh reinforcement topped with a 3mm decorative layer. Acrylic renders, consisting of two coats of 4 to 6mm reinforced with mineral fibre or glass mesh, are the next best performing, and are likewise highly elastic, weather resistant and breathable.

Concrete

A major disadvantage of concrete is its large carbon footprint: one tonne of cement resulting in the emission of approximately one tonne of CO_2. Furthermore, only 50 per cent is currently recycled for use in new building projects (compared to up to 99 per cent for structural steel). It is estimated by BRE's Green Guide that 50 per cent of concrete is crushed and recycled, 40 per cent is down-cycled for use such as hardcore in substructure works or road construction, and the remaining 10 per cent is waste that goes to landfill. Down-cycling does help to reduce the use of aggregates, but does not help to reduce the supply of materials needed for new concrete.

Steel

Steel is a robust and long-lasting material, despite being moderately energy intensive, and it can be easily purchased as a recycled product. Over 85 per cent of steel is recycled at the end of its life. In UK construction, the reuse and recycling rates of various steel products have been estimated at 92 per cent for rebar, 85 per cent for hot-dip galvanised sheet and 99 per cent for structural steel sections. By saving remelting, reuse is the most environmentally advantageous approach at the end of a building's life. The energy used in producing steel from recycled steel is roughly one-third of that for new steel. Recycling steel saves energy, CO_2 and resources by displacing the need to make more steel from virgin sources.

Cement render is liable to crack. Polymer modified cementitious render, developed to combat this problem, is lighter, but does not perform as well as the other options.

Internal insulation

The most important piece of general advice is to avoid covering cables with insulation, since this can cause them to overheat and create a fire hazard. PVC sealing on electrical cables may degrade in contact with polystyrene, so it is important to use cover strips or to place them in ducts. Cables less than 50mm from the surface of the plasterboard should be enclosed in metal conduits. Where ceilings, floors and internal walls join the main outside walls there will be thermal bridges. It is a good idea to return insulated dry lining a short distance along these internal services to avoid the risk of condensation.

There are two techniques for internal insulation: using insulated plasterboard applied directly on to the internal walls, or using studwork. The first of these techniques will achieve a higher insulation the greater the depth. Airtightness is preserved by applying a continuous adhesive strip to it around the edges of the

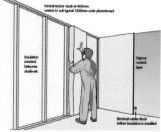

Figure 4.17a Installing internal insulation, dry lining style.

Source: EST

Figure 4.17b Installing internal insulation by fixing it directly to the wall.

Source: EST

wall and all openings such as sockets and plumbing. The best boards include a vapour control layer to stop moist internal air condensing on the cold break behind the insulation.

The second technique is employed on a wall that has previously suffered from damp. It is also recommended where the surface is not in a strict plane. The timber used as studwork acts as a relative thermal bridge, which must be broken by insulation on the front. Alternatively, insulating studs, which are made of extruded polystyrene laminated to oriented strand board, are available. A damp-proof membrane is placed between the studs and the wall. Insulation is placed between the studs. Insulation should be taken into window and door reveals.

An intelligent vapour control layer is fitted over this by continuously lapping and bonding the membranes together, sealing them back to the floor, internal walls and windows. This acts as the airtightness barrier. At all costs steps must be taken to avoid puncturing this barrier. Occupiers must be informed of this.

Floors

In the case of a solid ground floor, insulation should be applied over the top. Beneath this the vapour control layer is applied, which laps up the sides all around to join the one coming down the wall. The floor surface is laid on top of the insulation.

If the floor is being replaced, then, from the bottom up, the levels are: sand, vapour control or damp-proof layer (which again laps up the sides to join the one coming down the wall), insulation, then either concrete plus tiles, or timber board flooring. If the floor is being replaced, the installation of underfloor heating is greatly recommended, as comfort levels are raised and heating costs lowered (see Chapter 7).

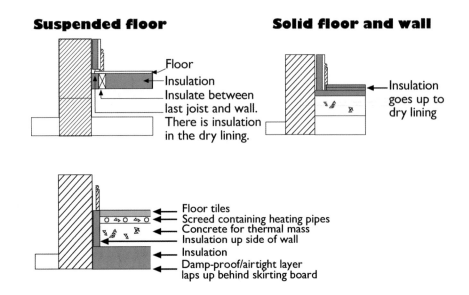

Suspended floor

Solid floor and wall

Floor
Insulation
Insulate between
last joist and wall.
There is insulation
in the dry lining.

Insulation
goes up to
dry lining

Floor tiles
Screed containing heating pipes
Concrete for thermal mass
Insulation up side of wall
Insulation
Damp-proof/airtight layer
laps up behind skirting board

Roofs

The type of roofing insulation depends on the type of roof: flat or pitched. External wall insulation may be carried over a flat roof. A U-value of at least 0.25W/m²K should be aimed for. The insulation in a flat roof should ideally be located between the roof deck and the weatherproof membrane in a warm roof deck construction. Careful detailing at the edge and parapet areas of flat roofs is vital for reliability and durability.

For a pitched roof the aim is 0.16W/m²K. Here, it can only be installed externally if the roof is being replaced. Otherwise, internal roof insulation is applied in a similar way to dry lining on vertical walls, with studs. On the inside, fix foil-faced laminated insulating plasterboard over the existing ceiling, or plasterboard over quilt insulation over interlocking foil-faced boards. There must be a ventilation path above the insulation at least 50mm or two inches deep. There must be absolutely no gaps between the slabs. Use caulking or firmly applied tape on the mechanically clamped joints between the two solid plates, for example, between a stud and plasterboard.

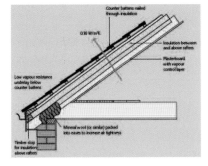

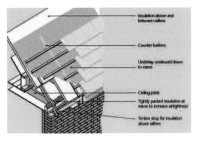

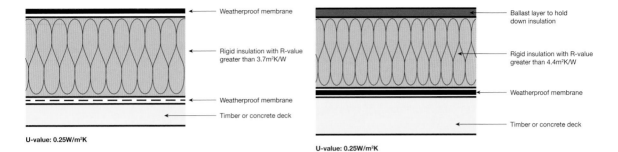

Weatherproof membrane

Rigid insulation with R-value greater than 3.7m²K/W

Weatherproof membrane

Timber or concrete deck

U-value: 0.25W/m²K

Ballast layer to hold down insulation

Rigid insulation with R-value greater than 4.4m²K/W

Weatherproof membrane

Timber or concrete deck

U-value: 0.25W/m²K

At the base of the sloping ceilings, purpose-made eave vents should be installed that provide the equivalent of a 25mm or one inch continuous gap, as well as ventilation at the roof's ridge in order to cross-ventilate the roof space and prevent condensation. The loft hatch should be insulated and draught-sealed or a proprietary insulated access hatch fitted.

Figures 4.20a and b
Warm roof (a) and cold roof (b) flat roof insulation.

Source: EST

Passivhaus Standard

The Passivhaus Standard (see www.passiv.de) has been developed over the past 20 years for new and existing buildings, and is being confirmed as the safest method for achieving low and zero carbon buildings. It stipulates a way of analysing all of the heat gains and losses within a building, and modelling improvements through the use of software to achieve the optimum result cost efficiently. Various established strategies are advocated for achieving the standard, which is defined in terms of the energy used by the building.

The standard requires that the space heat requirement should not exceed 15kWh/(m²/a). Total primary energy use (of all appliances, lighting, ventilation, pumps, hot water) must also not exceed 120kWh/(m²/a) (38039Btu/ft²/yr). Building fabric U-values must be less than 0.15W/m²K. This implies an airtightness level of 0.6 times the building volume per hour at 50 Pa (N/m²), as mentioned on p. 65. Even if a refurbished building is not certified as Passivhaus, it could still achieve the EnerPHit Standard, which is designed for refurbishing. With this, the criteria have been relaxed a little so that the space heating energy requirement is 25kWh/m²/yr, rather than the 15 required for Passivhaus, with an airtightness level of one building volume air change per hour.

The strategies available address thermal bridging and airtightness and all energy use within the building. Designs are modelled using the Passivhaus Planning Package (PHPP) software, which may be purchased from the Passivhaus website. Thousands of new buildings have been built to this standard and it is rapidly gaining credence internationally owing to the simplicity and flexibility of its approach compared, say, to national Building Regulations. Although many of these are dwellings, they include blocks of flats and office buildings and, in principle, most building functions can be catered for.

U-values for windows and doors generally need to be less than 0.8W/m²K (4755 Btu/ft²/yr) (for both the frame and glazing) with solar heat gain coefficients around 50 per cent. BRE, a statutory body in the UK promoting Passivhaus, provides Solar Heat Gain Coefficients (SHGC, which is defined as the proportion of solar energy that enters via the window) to help with calculations that may be adjusted for glazing on different façades. This can help either reduce heat loss on sheltered sides/ north-facing glazing, or reduce the likelihood of overheating when specified in conjunction with other features/strategies (the SHGC of a window usually decreases as the U-value improves).

The requirements imply the following features to achieve them:

- Passive preheating of fresh air: brought in through underground ducts that exchange heat with the ground to reach above 5°C (41°F), even on cold winter days (see Chapter 6);
- MVHR (Mechanical Ventilation with Heat Recovery from the expelled air transferred to the incoming air): transfers over 80 per cent of the heat in the ventilated exhaust air to the incoming fresh air (see Chapter 7);
- Hot water supply using renewable energy: solar collectors, biomass, CHP or heat pumps powered by renewable electricity (see chapters 6 and 8);
- Energy-saving appliances: ultra-low energy lighting, refrigerators, stoves, freezers, washers, dryers and so on.

Case study: Staunton-on-Wye Endowed Primary School in Herefordshire, England

Figure 4.21 The completed, superinsulated Staunton-on-Wye Endowed Primary School building.

Source: Staunton-on-Wye Primary School, England

This may not be certified Passivhaus, but it is certainly an eco-minimal building, originally built in the nineteenth century. A newly completed rebuild for 104 children and 24 staff includes high standards of insulation throughout, a sedum roof, beneficial solar gain and good levels of daylighting. The designers avoided materials and treatments high in volatile organic compounds, and inert construction materials were used as much as possible.

The three primary school classrooms face south to optimise on daylighting, while the offices and main hall face predominantly north. The preschool is a separate building. Both share a 45kW woodchip-fired heating system. This is controlled by a Trend Building Management System (BMS), with monitoring, energy metering and temperature control. It serves a low-pressure hot water (LPHW) underfloor heating network. Small radiators have been installed in some small rooms with a higher heat loss. Room temperatures are controlled by the BMS, with a 2-degree band of adjustment via room thermostats. The school hall is heated by a three-speed low-pressure water fan convector. There is also a 3.9kW photovoltaics system.

The building fabric is a form of breathing wall construction, with timber frame sandwiching 250–300mm layers of Warmcel (recycled newsprint) cellulose insulation. A bituminous fibre board serves as the external covering, with oriented strand board and taped joints as the internal airtightness layer. This led to a U-value of 0.12, while the roof achieved 0.1.

The building has an airtightness average of $1.7m^3$ $(h.m^2)$ at 50Pa, above the Passivhaus standard but way below Building Regulations. The designers have provided bulkheads in ceilings and notional ductwork space for the retrofitting of a mechanical ventilation and heat recovery system. The building is naturally ventilated, single sided for the offices and cross-ventilation in the classrooms, where the ventilation is controlled by manual Teleflex winders for the fanlights on the south façade and for the extract through high-level clerestory windows. Automatic motorised windows under BMS control were considered, but not only was this too costly, full manual control of the windows was preferred by the teachers. Only the high-level vents in the hall are motorised.

Lighting controls comprise presence detection (auto-on and -off) for circulation areas, and manual switching with daylight-linked dimming to a preset level, and presence detection (auto-off) for the high-frequency fluorescent fittings in teaching areas.

A rough calculation indicates that the building has a space heating demand of around $29kWh/m^2$ per annum (with MVHR). This is almost double the Passivhaus standard of $15kWh/m^2$ per annum, but still twice as good as most building regulations require, and which is what the heat demand (estimated at 65 kWh/m^2 per annum) would be without MVHR recapturing the waste heat. This new building will be subjected to a two-year building performance evaluation. Interestingly, an error has already been found. All the incoming utility meters report back to the BMS, as do the sub-meters for the electrical distribution boards. Lighting and power is metered separately. It has been found that there is a 13 per cent disparity between the sub-meter totals and the main meter reading. It is being investigated to find out whether one or more sub-meters are reading inaccurately, or whether an electrical load is not being metered.

All passively designed buildings deserve to win their plaudits based on proven performance in use rather than through a post-completion calculation.

Enhanced construction details

It is worth mentioning a free tool that is very useful in some circumstances. Enhanced construction details (ECDs) that focus on heat losses that occur at the junctions between building elements (walls, ceilings, floors) and around openings are freely available from Britain's Energy Saving Trust. They are designed to help the construction industry achieve high performance standards and were developed in association with an industry working group. Using the complete set of three ECDs and ensuring that all remaining details achieve regular standards will obtain a thermal bridging y-value of $0.04 W/m^2K$. See http://bit.ly/Qsbs2r.

Note

1   John A. 'Skip' Laitner, Steven Nadel, R. Neal Elliott, Harvey Sachs and A. Siddiq Khan, The Long-Term Efficiency Potential, Washington, DC: American Council for an Energy-Efficient Economy, 2012 (http://www.aceee.org/research-report/e121).

Energy managers share their experience

Phil Bilyard, Energy Consultant, UK

What does your job involve?

I am responsible for reviewing historic energy data, undertaking energy audits of commercial building stock (operational buildings), benchmarking, reviewing load profiles of electricity and hydrocarbon fuels, identifying anomalies, undertaking energy surveys to quantify the audit findings and develop an appreciation of the sites' energy, occupancy and building use, focusing on opportunities for savings, developing a prioritised action plan (zero and low cost, capital investment requirements; short- , mid- and long-term), and producing technical reports on findings and targets for energy reduction.

I then agree with the client on the content and delivery. Where necessary, financial cases are prepared for any capital investment projects and we agree a contract term for the delivery of the incentives that have been identified.

Having set targets for reduction, monitoring activities are then commenced to confirm to the client that savings are being achieved, we are meeting targets, strategies and incentives are fulfilling objectives and routinely reviewed. This is an ongoing commitment. Any special project delivery work is also managed during this phase.

Generally, we support our clients with building engineering service advice and consultancy, monitor maintenance contractor performance and raise general awareness on standards and requirements.

What qualifications did you gain?

- Higher National Certificate in Electronic/Electrical Engineering (HNC);
- Higher National Diploma in Engineering (HND);
- Graduateship of the City & Guilds (GCGI);
- Certificate of competence in the fundamentals of energy managment (delivered under the TEMOL programme; Energy Institute and University of the West of England);
- Chartered Energy Manager (Energy Institute);
- Incorporated Engineer (IEng).

What does your day-to-day work look like?

No two days are the same, but normally site-based surveying, incentive implementation, undertaking performance audits, project management delivery, attending meetings or home-based compiling technical reports, energy modelling, monitoring and targeting activities, and dealing with escalations.

What do you love about your job?

Three-fold, saving energy for our clients, which pays for our costs and delivers additional savings, by default, doing our bit for the environment by reducing carbon footprints and earning a fairly reasonable salary. One of the most pleasing aspects of the job is seeing energy use drop consistently on a month-by-month basis and really demonstrating our worth to the client measured by our success: very satisfying.

What do you think are the biggest challenges?

There is a whole host of legislation surrounding new buildings, ensuring that energy conservation is a focus and, at the very least, consideration of renewable, low and zero carbon technologies; namely, the UK Building Regulations and in particular, Part L deals with energy conservation requirements and initiatives. This does extend to refurbishment work and extensions of existing buildings.

Frustratingly, operational commercial buildings offer a huge potential in energy savings, although this is often overlooked and not dealt with specifically in day-to-day operational aspects. The EU Energy Performance of Buildings Directive has brought about compulsory certification and reporting in the form of Energy Performance Certificates (EPCs), Air Conditioning Inspections and Display Energy Certificates (DECs) but apart from the latter, while providing a reasonable amount of information on opportunities for savings, the client is under no obligation to implement the recommendations. Most organisations see the documents as a compliance issue and put a tick in the box.

Little emphasis at national level is given to optimising building performance (i.e. using what you have, ensuring everything is operating efficiently, the building is being used appropriately and then measuring success and performance) before investing in technology and 'widgets'. When you consider the operational building stock, so much can be done with existing equipment and controls to save significant costs and reduce the carbon footprint with zero and low-cost incentives. More often than not, the simple measures are not being implemented.

Austerity measures in a depressed economic climate make it difficult to convince potential clients of the necessity, and that they should be budgeting for our consultancy fees, which will save them significant operating costs, reduce carbon footprints and fulfil aspects of Corporate Social Responsibility (CSR). Really, when energy management should be blossoming in a period such as this because of the direct cost benefits, it is still very difficult to get the message across. Principally, this is because our costs are a direct service line lumped in with other building operational costs, whereas the savings come from a utility budget, or, in the case of multi-tenanted buildings, the tenants often pay for their energy separately and as used, so savings are not realised until later. Some organisations place very little emphasis on energy management, and are concerned merely with ensuring that all utility costs are paid for regardless of how efficient the building may be operating or how much can be saved.

Sometimes the waters have been muddied before us with some organisations that are linked directly to the sale of products and technology. Often, a client is encouraged to spend, in some cases considerable funds, to install equipment that is inappropriate for a particular building or application and does not deliver to expectations. Of course technology has its place, but the basics need to be addressed first.

What are you most proud of?

Achieving Chartered Energy Manager and Incorporated Engineer (IEng) status and professional recognition.

What's the best way to engage people on energy efficiency?

First, one must have an understanding of the organisation you are working for, its idiosyncrasies and the structure. It's a multifaceted approach; you have maintainers, building users and building managment teams, and it is important that all departments are singing from the same hymn sheet; otherwise, if you intend to implement change and not all stakeholders have been involved in the process, any plan will have a limited impact.

So, first, the most important approach is inclusivity and engaging all stakeholders in the process. Raising awareness, attending briefings and training not only dispenses with necessary information to deliver success, but builds relationships, trust and motivation. There must be a clear and common objective among all parties; otherwise the plan will not succeed.

A top-down approach is necessary with briefings and meetings at board level, middle/line management and at end-user level. Any forum, whether it be at meetings, training events, staff events, etc., should be structured to encourage participation and be regular in order to keep the focus on energy management. Regular departmental meetings should always have an energy section to cater for the dissemination of information and also to record suggestions and ideas.

Of paramount importance is ownership. Each department or organisation should be represented, and individuals should have allocated responsibility for aspects of energy management. Cooperation, common objectives and ownership will develop an inter-organisational team spirit. The board of directors must be prepared to allow nominated personnel the time and resources to dispense with their often extracurricular duties.

Energy incentives, delivery and status should be regularly reported and, of course, the savings should be communicated boldly to develop pride and motivation. All targets should be realistic, because it can become demoralising if targets are not achieved and repeatedly fall short.

What is the first thing that you would recommend to save energy and carbon emissions?

Ensure that it is firmly understood where energy is being used, how it is being used and where inefficiencies exist. You can't save energy until you understand where and how much is being used. This may require additional metering in particularly complex buildings where there is limited information.

Focus on high energy use, review plant condition, operation and control. Optimise for system performance and energy efficiency before looking towards technology and new products.

What is your favourite tool of your trade?

Energy meters and bespoke modelling tools developed in Microsoft Excel. Energy simulation and modelling tools are also very useful.

5

Lighting

As a proportion of overall energy use, lighting can figure surprisingly large, being responsible for up to 30 to 40 per cent of an electricity bill in places which have not embraced energy-efficient lighting. No other single activity in a normal office building uses more energy, according to the US Energy Information Administration (EIA).

The EIA estimates that more floor space is lit in office buildings than any other type of commercial structure: over 10 billion square feet in 2009. Of course, it varies according to sector: schools might spend 20 per cent of their electricity budget on lighting and offices double this, whereas for heavy industry and manufacturing it will be a smaller proportion. The figures are the same in most developed countries. Nevertheless, there are usually plenty of opportunities for reducing the cost of lighting by up to 90 per cent (if daylighting is maximised and other lighting is converted to LEDs), while at the same time improving the quality of illumination for people occupying buildings.

Building regulations in different countries specify certain levels of lighting and efficiency for different purposes. In California, for example, the Energy

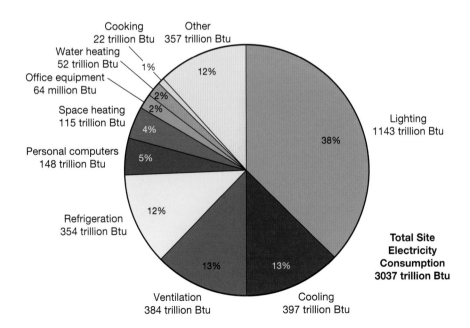

Figure 5.1 In 2009, lighting used more power than any other application in American industrial and commercial/office buildings.

Source: US Energy Information Administration

Figure 5.2 Efficiency gains for selected commercial equipment in three cases, 2040 (percent change from 2011 installed stock efficiency). Lighting clearly holds the most potential.

Source: EIA Annual Energy Outlook 2013

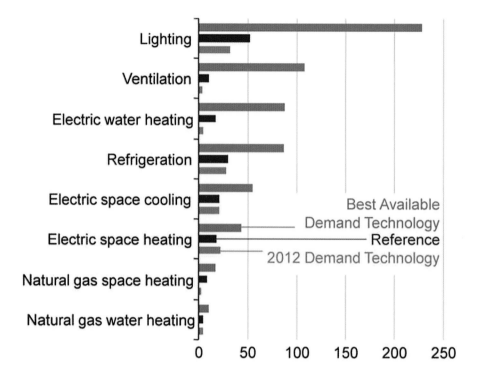

Independence and Security Act of 2007 requires that commercial building owners must ensure that lighting uses less than 1.1 watts per square foot. This means that light bulbs must have 25 per cent more efficacy than the standard 100-watt incandescent bulb nowadays. They will be just as bright (1600 lumens) but will consume no more than 72 watts. However, the best available technology is far more efficient than this, as we will see. The first priority is to maximize the free availability of sunlight. To do this we need to understand how much light is needed for a given space.

Light

Light is measured in lumens, and illumination in lux. The efficacy of a lamp is measured in the number of lumens it produces per watt input.

A lux (symbol: lx) is equal to an illumination level of 1 lumen per square metre. A lux reveals how many lumens are needed to light a given area. One lux is equal to 1 lumen per square metre. (In non-SI units, 1 foot-candle is equal to approximately 10 lux.)

Light intensity decreases by the square of the distance from the bulb. Therefore, 500 lux directed over 10 square metres will be dimmer than the same amount spread over 1 square metre. The strategy is to maximise the number of lumens obtainable for the least number of watts. If an office, which requires 500 lux, occupies an area of 100 square metres and the lamps are two metres above the desk level, this will need 500 × 100 × 4 = 200,000 lumens. This office could

Table 5.1 The performance of typical 12V lamps

Lamp type	Rated watts (W)	Light output lumens (lm)	Efficacy (lm/W)	Lifetime (hours)
Incandescent globe	15	135	9	1000
Incandescent globe	25	225	9	1000
Halogen globe	20	350	18	2000
Batten-type fluorescent (with ballast)	6	240	40	5000
Batten-type fluorescent (with ballast)	8	340	42	5000
Batten-type fluorescent (with ballast)	13	715	55	5000
PL-type fluorescent (with ballast)	7	315	45	10000
LED lamp (see note)	3	180	30–100	>50000

Source: Manufacturers' data

Note: The performance of LEDs varies considerably according to the manufacturer. Choosing the right LED products is very important.

Table 5.2 How many lux are needed for different applications

Lux level	Area or activity
20–30	Car parks, roadways
<100	Corridors, stores and warehouses, changing rooms and rest areas, bedrooms, bars
150	Stairs, escalators, loading bays
200	Washrooms, foyers, lounges, archives, dining rooms, assembly halls and plant rooms
300	Background lighting (e.g. IT office, packing, assembly (basic), filing, retail background, classrooms, assembly halls, foyers, gymnasium and swimming pools, general industry, working areas in warehouses)
500	General lighting (e.g. offices, laboratories, retail stores and supermarkets, counter areas, meeting rooms, general manufacturing, kitchens and lecture halls)
750	Detailed lighting (e.g. manufacturing and assembly (detail), paint spraying and inspection)
1000	Precision lighting (e.g. precision manufacturing, quality control, examination rooms)
1500	Fine precision lighting (e.g. jewellery, watchmaking, electronics and fine working)

Source: Carbon Trust and lighting manufacturer Veelite

Figures 5.3 and 5.4 Illumination decreases by the inverse square law with distance from the light source. As an example, if a bulb gives off 400 lux at 1m, at a distance of 4m the irradiation will be one-sixteenth of this, or 25 lux (reading from the graph). If 300 lux is required at 4m distance, then 12 lamps each giving off 400 lux would be required (300/25). This illustrates the importance of positioning in lighting.

Source: Author & Wiki Commons. Author: Borb

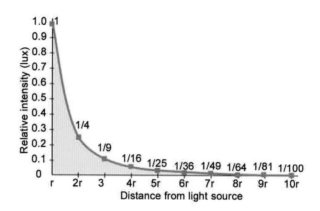

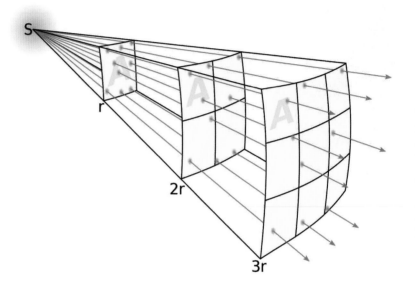

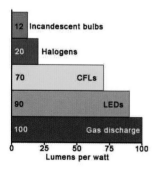

Figure 5.5 The average efficacy of different types of lighting.

Source: Author

therefore be lit (at night) by (using the figures in Table 5.2): 200,000/70 = 2156W of CFLs, 200,000/90 = 2,224W of LEDs, or 2000W of high-frequency fluorescent tubes.

Using natural light

Now let us consider daylighting. Visible transmittance (VT) is a fraction of the visible spectrum of sunlight (380 to 720 nanometers), weighted by the sensitivity of the human eye, that is transmitted through the glazing of a window, door, or skylight. A product with a higher VT transmits more visible light. VT is expressed as a number between 0 and 1.

This is different from the solar heat gain coefficient (SHGC), which is the proportion of total solar energy that enters via a window. Light-to-solar gain (LSG) is the ratio between the SHGC and VT. It provides a gauge of the relative efficiency of different glass or glazing types in transmitting

Figure 5.6 Daylighting should be maximised with the help of windows, but its management can be separated from that of heat gain, which may not always be required. Shading will help control unwanted glare.

Source: IEA

daylight while blocking heat gains. The higher the number, the more light is transmitted without adding excessive amounts of heat.

Building occupants feel more comfortable and will have a higher degree of well-being in sensitively lit interiors, which means maximising the use of natural light. Light directly affects mood and alertness as well as productivity, as borne witness by Seasonal Affective Disorder (SAD).

Passive daylighting

Even existing buildings can be adapted to make better use of natural lighting. In Chapter 4 we looked at some ways in which windows may be used to admit natural light without causing too much glare throughout the year to building occupants. Bright reflective surfaces and light surface colours will aid in the distribution of daylight in rooms, but this is by no means an exhaustive list.

Passive daylighting is a system of collecting sunlight using static, non-moving and non-tracking systems such as windows, light shelves, skylights, roof lights or atrium spaces, tubular daylight devices, solar shading devices, daylight-responsive electric lighting controls and daylight-optimised interior design (furniture, etc.). The intention is to direct low daylight high into a space (to reduce the likelihood of excessive brightness). Where possible, ceilings can be sloped to direct more light inward. It is vital to prevent direct daylight from reaching critical visual task areas, and so it needs to be filtered. Artificial light should be brought in gradually further within spaces, so that there is not a sudden contrast between natural and artificially lit areas.

Directing daylight into a building

Skylights are one way of achieving this. They can be either passive or active. Passive skylights simply let daylight enter through glazing in the roof. Active skylights contain a mirror system that tracks the sun across the sky to reflect and direct it where needed. Optionally, systems are available which reduce daylight during summer months and assist with cooling and ventilation.

A light shelf is an architectural element that allows daylight to penetrate deep into a building. They are effective on south façades but often ineffective on east or west elevations of buildings. A horizontal, light-reflecting overhang is placed above eye level and has a high-reflectance upper surface to reflect daylight on to the ceiling. They are commonly made of an extruded aluminium chassis system and aluminium composite panel surfaces. Light shelves and louvred systems make it possible for daylight to penetrate the space up to four times the distance between the floor and the top of the window. They are generally used in continental climates and not in tropical or desert climates due to the intense solar heat gain.

Exterior louvre systems, which are vertical or horizontal fins whose opening and shutting, rather like Venetian blinds, can be controlled automatically, prevent too much sunlight from reaching a window, or can have reflective surfaces to direct light inside while avoiding glare.

Architectural light wells direct daylight further into – mostly new – buildings. Light can be brought into spaces in other ways too: by using atria and light guidance systems. Light tubes, also called sun pipes, solar pipes or daylight pipes,

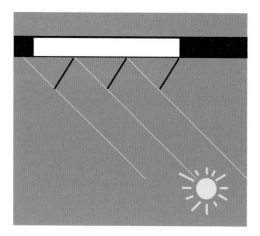

Figure 5.7 Plan view of a windowed wall showing adjustable louvre positioning on the outside to prevent glare or reflect deeper light into a building.

Source: Author

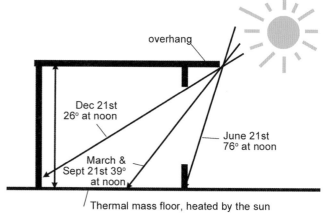

Figure 5.8 Cross section of a windowed wall showing use of an overhang to cut glare but maximise daylighting through the seasons.

Source: Author

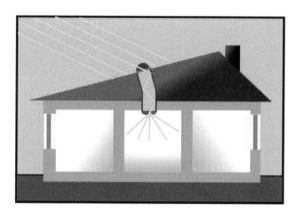

Figures 5.9a and b A sun tube or pipe fitted into a roof and taking light down to a windowless bathroom.

Source: (a) From German Wikipedia; (b) Sun tunnel image courtesy of Greenworks and Velux

have reflective inner surfaces to take light from a roof level deep into lower floors. Compared to conventional skylights and other windows, they offer better heat insulation properties and more flexibility for use in inner rooms, but less visual contact with the external environment.

Operable shading and insulation devices

A design with too much Equator-facing glass can result in excessive heating, or uncomfortably bright living spaces at certain times of the year, and excessive heat transfer on winter nights and summer days. Although the sun is at the same altitude six weeks before and after the solstice, the heating and cooling requirements before and after the solstice are significantly different. Variable cloud cover influences solar gain potential. This means that latitude-specific fixed window overhangs, while important, are not a complete seasonal solar gain control solution.

Control mechanisms (such as manual or motorised interior insulated drapes, shutters, exterior roll-down shade screens or retractable awnings) can compensate for differences caused by thermal lag or cloud cover, and help control daily/hourly solar gain requirement variations. Automated systems that monitor temperature, sunlight, time of day and room occupancy can precisely control motorised window shading and insulation devices. A successful lighting design that incorporates any of these architectural features must be integrated with the artificial lighting system using advanced lighting controls, and with the building energy management system (BEMS) if there is one. Three types of controls are available:

1   Those that turn lights off when there is ample daylight;
2   Control of individual lamps in zoned circuits that are progressively further from windows, to provide intermediate levels of light;
3   Dimming controls which continuously modulate the power to lamps to complement the amount of daylight.

Figure 5.10 This cross section illustrates various possible types of shading and light-directing fixtures for and around Equator-facing windows to maximise daylighting for interiors while controlling glare around the year and through the day.

Source: Author, adapted from Mr7uj

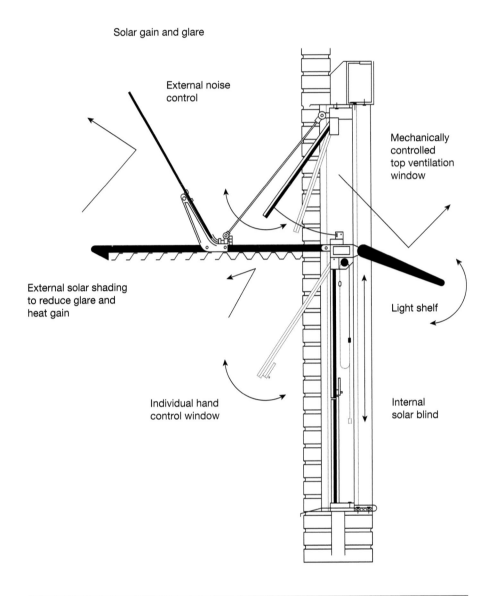

Solar gain and glare

External noise control

Mechanically controlled top ventilation window

External solar shading to reduce glare and heat gain

Light shelf

Individual hand control window

Internal solar blind

Case study: 41 Cooper Square, New York City

Number 41 Cooper Square was the first LEED-certified academic laboratory building in New York City, as verified by the Green Building Certification Institute. Pictured is the operable building skin, offset from a glass and aluminum window wall, which is made of perforated stainless steel panels. These panels reduce the impact of heat radiation during the summer, while in winter helping to insulate it. This solution may be applied as a retrofit cladding for existing buildings. Seventy-five per cent of this academic building's regularly occupied spaces are lit by

Figure 5.11 41 Cooper Square, New York City.

Source: Robzand

natural daylight, with the help of a full-height atrium that also improves the flow of air. Radiant heating and cooling ceiling panels contain HVAC technology that helps to make it 40 per cent more energy efficient than a standard building of its type. It is topped with a green roof that reduces city 'heat island' effect, and harvests rainwater for reuse. A cogeneration (CHP) plant provides power and heat.

Angle of incident radiation and shading coefficients

If the available year-round natural light within the building needs to be accurately modelled before refurbishment or new build, it becomes necessary to calculate the angle of incident radiation and shading coefficient.

The amount of solar gain and light transmitted through glass is affected by the angle of the incidence at which it hits the window. If this is within 20 degrees of the perpendicular it will effectively pass through it; at over 35 degrees the majority of the energy will bounce off. Optimum window designs for the available wall can be modelled using a photographic light meter and a heliodon (which adjusts the angle between a flat surface and a beam of light to match the angle between a horizontal plane at a specific latitude and the solar beam). This helps calculate the amount of light (and heat) energy entering the room based on the angle of incidence. Better still, software is available to perform the same job and

Figure 5.12 Heat and light are transmitted through glazing both directly (primary transmittance) and by being absorbed and then re-emitted by the glass via convection and radiation (secondary transmittance). The degree to which this is successful is called the 'solar heat gain coefficient' (SHGC).

Source: Author

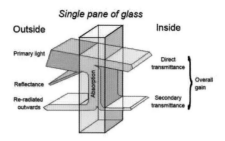

calculate cooling and heating degree days (see Chapter 7), and therefore energy performance, using regional climatic conditions available from local weather services.

Shading coefficients are used to measure the solar energy transmittance through windows. In Europe this coefficient is called the 'G-value' while in North America it is 'solar heat gain coefficient' (SHGC). G-values and SHGC values range from 0 to 1, a lower value representing less solar gain. Shading coefficient values are calculated using the sum of the primary solar transmittance (T-value) and the secondary transmittance. Primary transmittance is the fraction of solar radiation that directly enters a building through a window compared to the total amount of solar radiation (insolation) that the window receives. The secondary transmittance is the fraction of inwardly flowing solar energy absorbed in the window (or shading device) again compared to the total solar insolation.

Windows

Windows consist of two or three layers of glazing. Glazing is a pane of glass made specifically for this purpose by floating molten glass on a bed of molten metal to give it a uniform thickness. Once the desired thickness is obtained, it is annealed for protection, and may be coated or laminated to improve its strength or optical and insulation properties.

Coatings

This means that when specifying glazing or windows, it is possible to order them with different coatings that not only admit and retain infrared heat, but also admit as much light as is required. Manufacturers supply a huge range of coatings: some will permit only 6 per cent of light to enter the building, or 8 per cent of heat. Panes specifically designed to reduce energy use have extra clear outer layers, letting in up to 80 per cent of light and 71 per cent of the sun's heat.

'Smart' windows are also becoming available, based on electrochromism, a technology that can control light and heat while maintaining view and reducing glare. The glass can be clear, opaque, tinted or coloured, has the capability to modulate heat and light transmission, and may be used in a variety of applications. The windows' optical properties vary when an electric field or current is applied across the device. This causes the absorbence or the reflectance of the active layers to change, thereby modulating the amount of light (electromagnetic radiation) that passes through the coated substrates or glass.

Energy labels

Modern windows are rated by national bodies and come with a declaratory label. In the UK this is the British Fenestration Rating Council (BFRC). This label displays the following information:

1 the rating level: A, B, C, etc.;
2 the energy rating, e.g. -3kWh/(m²K)/yr (= a loss of three kilowatt-hours per square metre per year);
3 the U-value, e.g. 1.4W/(m²K);
4 the effective heat loss due to air penetration as L, e.g. 0.01W/(m²K);
5 the solar heat gain G-value, e.g. 0.43 (see below).

The ratings are ranked according to the amount of thermal transmittance (heat transfer) that is permitted by the window, as measured by the amount of energy (in kilowatt-hours) lost per year through 1 square metre of the window, divided by the difference in temperature between the inside and outside (kWh/(m²K)/yr):

Table 5.3 Meaning of the BFRC window energy label rating

Rating	Energy lost per year (kWh/m²K)
A	0 (no energy lost) or better
B	0 to −10
C	−10 to −20
D	−20 to −30
E	−30 to −50
F	−50 to −70
G	−70 or worse

Figure 5.13 A sample window rating label from the National Fenestration Rating Council (NFRC). It gives the solar heat gain coefficient (SHGC) for the glazing, the U-factor (the same as the R-value, or inverse of the U-value), air leakage rate and its visible transmittance (VT).

Source: NFRC

Figure 5.14 Another sample American window efficiency rating label.

Source: ENERGY STAR

In the USA, the equivalent label is produced by the National Fenestration Rating Council (NFRC). Its Component Modeling Approach (CMA) Product Certification Program enables whole product energy performance ratings for commercial (non-residential) projects. It uses the following components:

- Glazing: Glazing optical spectral and thermal data from the International Glazing Database (IGDB);
- Frame: Thermal performance data of frame cross sections;
- Spacer: Keff (effective conductivity) of spacer component geometry and materials.

Most US states' building energy codes reference NFRC 100 and 200 for fenestration U-factor and SHGC because they are required by ASHRAE 90.1, Section 5.8.2. California's Title 24 Building Energy Efficiency Standard now requires CMA label certificates for site-built fenestration in large projects.

Obtaining the label certificates is voluntary. The CMA program does not label each product. Instead, the NFRC lists the values for rated products on a single document for the entire project, called the label certificate. The CMA may be used to rate commercial windows, doors, skylights, curtain walls and storefronts. To obtain a label certificate, the design team uses pre-approved, NFRC-rated components (frames, glazing and spacers) to configure a product in the CMA Software Tool (CMAST), which then generates energy performance ratings for the whole product. An NFRC-Approved Calculation Entity (ACE) then certifies the ratings and issues the label certificate.

Lamps

From maximising daylighting, we move to the choice of lamp, which is to say, light bulb, together with its holder and shade or cover. Lamps are evaluated according to their efficiency, lifetime and colour. There are four types of lamp.

Incandescent

The oldest type of light, being phased out in many areas because of their inefficiency, achieves its brightness by passing an electric current through a wire, typically made of tungsten. Some incandescent lamps are slightly more efficient thanks to the use of halogen as a gas surrounding the tungsten element of the bulb. Halogen lamps are around 30 per cent more efficient than the old-fashioned type. They have good colour-rendering ability, but do not last very long (see Table 5.1).

Colour rendering and temperature

Nowadays, LED and CFL lights are available in a wide range of colour temperatures. Colour temperature is a way of describing how cool or warm the colour of lighting is. It is measured in degrees Kelvin, just like normal temperature. This may be found on the packaging. The lower the temperature, the redder, or warmer, the colour. The higher the temperature, the more blue, or cool, the colour. For lamps, Table 5.4 gives a rough guide.

Table 5.4 Colour temperature and perceived colour

Temperature	Perceived colour
2700K	very warm, yellow white
4000K	neutral white
6400K	daylight
8000K	sky white, almost blue

Colour rendering relates to the way objects appear under a given light source. The measure is called the 'colour rendering index' (CRI). A low CRI indicates that objects may appear unnatural under the source, while a light with a high CRI rating will allow an object's colours to appear more natural. For lights with a 'warm' colour temperature the reference point is an incandescent light. For lights with a cool colour temperature the reference is daylight. Table 5.5 lists typical colour rendering index ratings for a variety of lights.

Table 5.5 Typical colour rendering index ratings for a variety of lights

CRI	Lighting type	Application
22	High-pressure sodium lighting	Street lighting
62	Common fluorescent tube	Office
80–85	Compact fluorescent lighting (warm white)	Residential
85	Premium 4-foot fluorescent tube	Retail
80–90	Solid state LED lighting	Residential
95	Incandescent light bulb	Residential

Figure 5.15 Halogen lights.

Source: Hong Kong Lights

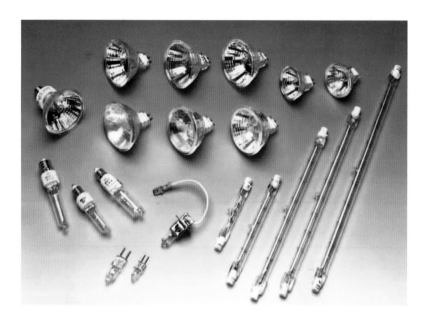

Fluorescent

This category includes fluorescent tubes and compact fluorescent tubes (CFLs). These work by passing the electric current through a gas which causes it to glow. The gas needs to warm up before it can shine, but with these types of lights it usually doesn't take more than one second. They have good colour-rendering ability. Their lifetime varies from 6000 hours for compact fluorescents, to 12,000 hours for fluorescent tubes. Long-life tubular versions exist which can last up to 70,000 hours. They can be switched on and off easily and are dimmable.

Figure 5.16 CFL bulbs are now available in many fittings, sizes and colour temperatures. Over their lifetime they can save up to 40 times the energy of a single incandescent equivalent, being eight times more efficient and lasting five times longer.

Source: Osram

CFLs now come in many different fittings and styles, including downlights, spotlights, dimmable wall lights, mirrors and so on. Many require 'ballasts'. Most dimmable ballasts require additional wiring – but some are available which do not – or changing a centre pendant. For multiple lights, the recall and earth cable is required to carry a permanent live cable and a switch.

LEDs

Solid-state lighting is divided into Light Emitting Diodes (LEDs) and Organic Light Emitting Diodes (OLEDs). LEDs have a very long life, typically over 50,000 hours, and emit a point source of light. They are often integrated into the light fixture, so there is no lamp to replace.

LEDs can last up to 2.5 times longer than CFLs, and 25 times longer than incandescents. This makes them ideal especially for inaccessible places, since they do not need to be changed very often. Their light is directed; therefore groupings of LEDs are required to obtain the equivalent amounts of lux to CFLs, and arrays of these pointing in different directions are needed in order to extend the angle over which their light can fall. Modern designs imitate old-fashioned bulbs to achieve omnidirectional effects, and are available in fittings which match those of halogen and incandescent lights. Where a halogen light of this type might consume 40W, the LED equivalent will consume between 1 and 5W. LEDs now come in a full range of colour rendering and temperatures.

Figures 5.17a–d LEDs are available in many colours and fittings, including, as may be seen, LED substitutes for fluorescent (in this case T8) and conventional screw fittings, and are eight times more efficient than halogen lights, while lasting 25 times longer.

Source: (a) Crown Copyright; (b) Rayco, used with permission; (c and d) GEB Lighting, used with permission

Figure 5.18 6000K LED street lights.

Source: Author

LEDs are now used in thousands of indoor and outdoor settings, replacing most uses for lighting, from street lights and traffic lights to space lighting, fridges and mood lighting.

Organic light emitting diodes (OLEDs) are a flat panel giving even, diffuse light, and are commonly found in consumer electronics such as mobile phones, computer monitors and computer notebooks. They use a film of organic compound, which emits light in response to an electric current.

Figure 5.19 LED lighting used in fridges in place of T5 fluorescent tubes.

Source: SunRay Lighting, used with permission

Gas discharge

Although fluorescent bulbs are also gas discharge, to make a distinction here this category includes sulphur lamps, metal halide and sodium lamps. They work by sending an electrical discharge through an ionised gas, or plasma. They offer long life and high efficiency, but are more complicated to manufacture and therefore expensive, and they require auxiliary electronic equipment such as ballasts to control current flow through the gas.

High-pressure sodium lights are used for warehouses and floodlighting. Bright orange, they have poor colour rendering and take a long time to achieve full brightness. They are efficient, at 125lm per watt, and last around 20,000 hours. They can be dimmed to a limited amount.

High-pressure mercury lights produce white light with a bluish tinge, consequently possessing poor colour rendering. They should be phased out. Metal halide and ceramic metal halide lamps offer an efficiency of 80lm per watt and a lifetime of 12,000 hours. Their colour rendering varies from Ra60 to Ra90 and they are available in sizes from 20W to 2kW. They are often used in shops, for exterior lighting and in sports grounds.

Luminaires

A luminaire is the fixture that holds the lamp. There is now a huge range of types of luminaire, but they all contain five main components: the housing, control gear, lamp holder, lamp and reflector. Reflectors direct light, maximise its usage, and thus reduce the quantity of light needed. Some luminaires may also include a diffuser.

The efficiency of a luminaire is measured by their light output ratio (LOR). The higher this is, the better. An important characteristic for efficiency purposes is the reflectiveness of the material used. Satin chrome reflects only half the light, whereas aluminium coated with silver reflects 90 per cent of it. Luminaires in offices should be designed to avoid glare on to screens. This type should not be used in retail environments, which have different requirements.

Lighting controls

Control gear is used to preserve lamp life for gas discharge lighting and, in addition, will provide manual or automatic dimming and switching. Modern controls use high-frequency electronic control equipment; this produces more light with less power. However, not all high-frequency lighting can be dimmed.

- Automatic and manual controls should be combined in areas where people need to control their own level of lighting. This means that users should be able to dim, and switch off and on lighting as they require, but lighting may also be switched off or dimmed automatically when not needed.
- Dimming gives occupants more control over their light levels and still realises savings.
- Adjustable light-level sensors can automatically turn a light on and off in response to changing amounts of daylight. These can be used outside, such

as in car parks or streets, or inside, to maintain an even level of light as the light from outside changes during the day.

- Occupancy sensors can tell whether a space is occupied and control lighting accordingly. They are appropriate for internal or external lighting, and come in several types:

  - Doppler sensors work by sending out high-frequency sound waves and listening for the bounce-back. When it is returned at a different frequency it knows there is a moving object around. It then sends a signal to the dimmable ballast to raise the light levels. When no movement is detected after a certain period of time, the light will return to its original level.
  - Passive infrared (PIR) motion sensors are used for many purposes, e.g. hand dryers and taps, but in the case of lighting can be vulnerable to dust and blocking objects, can be confused by radiators and fires, and have a shorter lifespan.
  - In some circumstances the use of a manual switch with a timer built into it may be appropriate.

Most control systems need additional connectivity. Nowadays, wireless technology is often cheaper to install than wired connections, as described in Chapter 1. Digital Addressable Lighting Interfaces (DALI) is one global standard used for intelligent lighting management. It is a protocol set out in IEC 60929 and IEC 62386, which are technical standards for network-based systems that control lighting in buildings. A DALI network consists of a controller and one or more lighting devices (e.g. electrical ballasts and dimmers). Each device is assigned a unique static address, allowing it to be remotely controlled. DALI also attempts to reduce the standby parasitic power losses of control equipment.

Upgrading halogen lamps and fluorescent tubes

A range of products incorporating LEDs is now available to replace many display and directional lamps, especially tungsten halogen, but also fluorescent tubes.

Figure 5.20 This 4 × 14 watt T5 600 × 600 modular fitting will replace the less energy-efficient 4 × 18-watt T8 switch start recessed fittings often found in older installations. The combination of high-frequency control gear and the long-life high-output 14-watt T5 tube can result in 25 per cent lower running costs, by comparison.

Source: SCL Direct

They combine a light source, power supplier, optics and heat management components. They can be more expensive but last at least 10 times longer than the incandescent lamp they frequently replace. It is important to check before purchase if dimming functions are required that the unit is compatible.

Retrofit kits are available to convert the less efficient non-high-frequency fluorescent light fittings to use T5 fluorescent or LED lamps. (T indicates that the shape of the bulb is tubular, and the last number (y) is the diameter in eighths of an inch.) High-frequency control gear uses around 10 per cent less electricity than the mains frequency equivalent, improves lamp life and eliminates flicker; with mains frequency control gear, for T12 or T8 tubes, lamps often flicker as they switch on. The LED replacements come in the form of an array, whose fittings match those of the tubes. If the lighting is over ten years old, the fittings should be replaced. This is an opportunity to install lighting controls.

Case study: Three Horse Shoes, Llandovery, Wales

Wyn Morgan's take-away and eat-in restaurant business was lit by 105 50W GU10 dichroic spotlights. These consumed 4.2kW for 12 hours per day, yielding a typical cost including electricity, purchase and replacement time in 2011 of over £1000 per month.

To try to reduce this expense he had started buying the best low-energy light bulbs that he could find locally: 11W Megaman 300K Megawhite CFLs. They cost £11 each and were rated for 2000 hours or six months. They reduced his lighting load to 1.1kW, but they did not last nearly as long as expected. Investigation showed that the lack of a heat sink in the bulb was causing them to reach 105°C/221°F and burn out. The manager was visiting a store to buy ten replacement bulbs each month. With 105 units it was impossible to know if each one had lasted the predicted 2000 hours. In addition, the inset units were difficult to grip when removing them; quite often the glass would detach from the unit, sometimes fragmenting and dropping toxic shards of glass over the restaurant floor. Invariably, failed units cast a shadow over the premises.

Following advice, Morgan replaced all of the light bulbs with Toshiba 8.5W 3000K Gu10 LED bulbs. These were five times more expensive but rated to last 40,000 hours (ten years), or 20 times longer, and were guaranteed for five years. Their ceramic heat sinks dissipated the heat, ensuring their longevity. A contractor, Dali Lites in Milton Keynes, financed the replacement with a 'Pay As You Save' loan that involved paying for all 105 bulbs over 18 months. The business would be saving money from the moment the lights were switched on. In addition, the restaurant looks much better. The cost of the bulbs up front would have been £2640. The energy, cost and time saving of not having to replace the units are together saving the business an estimated £3000 per annum.

Estimating the payback period for lighting replacement

1 Find the total power use by adding the power rating of all fluorescent lamps to the power consumption of the control gear (A).
2 Do the same for the new fluorescent or LED lamps and control gear (B).
3 Subtract B from A. Convert to kilowatts (kW) (C).
4 Estimate the annual operating hours of the system (D).
5 Multiply C by D to give the annual electricity usage (E).
6 Multiply E by your current electricity cost (F).
7 Obtain an estimate of capital cost and time cost of the replacement equipment (G).
8 Divide G by F to find the simple payback period in years (not accounting for inflation) (H).

Example

A room containing 30 100W T12 fluorescent lamps, not using high-frequency control gear, is to be replaced with T5 lamps using retrofit kits costing £40 each, including installation. The lights are typically on from 9 a.m. to 5 p.m. Monday to Friday. Electricity costs 13p/kWh.

1 $A = 30 \times 100W + 13W$ [control gear] $= 3{,}013W$.
2 $B = 30 \times 35W + 4W = 1054W$.
3 $C = A - B = 3{,}013W - 1054W = 1959W = 1.959kW$.
4 $D = 8$ hours per day $\times 5$ days $\times 52$ weeks $= 2080$ hours.
5 $E = C \times D = 1.959kW \times 2080$ hours $= 4{,}074.72kWh$ per year.
6 $F = 13p \times 4074.72 = £529.71$ per year.
7 $G = 30 \times £40 = £1200$ (labour cost may need to be added).
8 $H = G/F = £1200/£529.71 = 2.265 =$ years of operation to pay back the replacement cost.

Case study: The Springfield Literacy Center, Pennsylvania, USA

Designers unfailingly find that using modelling techniques at the design stage can produce huge savings over the lifetime of a building, at the same time as avoiding huge mistakes. Several software applications are available for this purpose. This does not mean that modelling always produces accurate figures, however, as installation standards and occupant behaviour have a part to play, as we note in the Conclusion. Nevertheless, this case study illustrates the benefit of such an approach.

Figure 5.21
The Springfield
Literacy Center
in Pennsylvania
after completion.

Source: IES

The Springfield Literacy Center in Pennsylvania wanted to achieve LEED certification. However, the site suffered from poor orientation and a nearby shade-causing woodland which dictated a split form for the building design and increased depth for the classrooms. There was a potential trade-off between achieving the LEED Daylighting Credit (IEQ c8.1) and the LEED Energy Use Credit (EAc1 – Optimize Energy Performance). Architects Burt Hill used software to conduct preliminary analysis on different components of the building envelope to quantify their influence on daylighting and energy performance.

The glazing size, type and shape options were analysed, together with the influence of the wall assemblies and natural ventilation options. This determined the size and location of the windows to achieve the most effective daylight levels for the students' tasks. The original design incorporated a bank of floor-to-ceiling windows in each classroom, but it was discovered that this would lead to too much glare. A section of window at floor level was removed and smaller windows inserted higher up. External shading devices and lightshelves were also incorporated.

The thermal performance of glazing specifications was assessed to balance the desire for transparency with energy efficiency. This required a higher U-value, based on winter exposure. A final check found there was still an incomplete resolution of the conflict. Therefore the software was employed to look at other daylighting parameters outside the LEED requirement, and a higher energy efficiency for the LEED Optimize Energy Performance credit was found to be possible by increasing the level of insulation in the wall (higher R-value). Thermal analysis looked thoroughly at heat losses and gains in order to avoid summertime overheating. The final wall composition was cavity brick with a U-value of 0.065. For this climate zone (6), the maximum allowed by the Code is 0.084.

'When the architectural design model is also the energy model, energy modelling and responsiveness become core components of the design process. For example, rooms are designed explicitly as 3D energy-consuming volumes instead of just floor space bounded by walls,' said Dustin Eplee, leader of Burt Hill's Energy Analysis team.

Energy managers share their experience

Andrew Bray, Director of Energy Management at da Vinci Construction

What does your job involve?

I am directing staff within my department and liaising directly with clients.

What is the first thing that you would recommend to save energy and carbon emissions?

Go for the quick wins. Whether this is changing to low-energy light bulbs, turning them stats down a notch, or spending half an hour reprogramming the BEMS, which might save tens of thousands of pounds. Far from these easy wins already having been achieved, we are, in every organisation I have been into, a country mile away from getting them.

As a preliminary exercise I always use the matrix [in Chapter 3]. For example, working with a client Deloitts in Amsterdam, the Netherlands, ten minutes spent filling out this form brought about an interesting debate about where they are and where they need to be, which resulted in the generation of a realistic strategy that ended up saving tens of thousands of pounds.

The exercise was repeated with Impress manufacturing company and many others. It is not an exact science, since it is up to individuals themselves to estimate where they are on the scale. Instead, it is a tool for self-assessment under which targets can be set.

The next step is to publicise what you have done. It's important to do this both internally and externally, so that you can build up support for everything that you do.

I am a firm believer in asking clients to improve the building fabric first. It's no good putting on solar panels if you haven't done this.

What do you think are the biggest challenges?

I often find that there is a conflict between the energy manager and the facilities manager. The maintenance people can block the changes you want to implement, or employees can complain they are too hot or too cold. Facilities managers have their key performance indicators to attain; they do not have it in their job description to be energy efficient. Building Energy Management Software can conflict with all of these and then the engineer will give you the runaround.

In such a situation it is necessary to take it to the account director, to go above them and get a mandate from management for cooperation from all of the staff. It's important to lay the groundwork by having a properly thought-out energy-saving strategy that has been signed off at the top level. Having management support means that energy managers can deal with opposition.

A shared-savings model, where both we and the client benefit from savings made, focuses their minds, and means you can attain the easy wins quickly.

Certain kinds of projects, like PFI (private finance initiative) contracts, can actually have clauses in them that mitigate against energy efficiency. I have had people threaten a breach of contract for people turning off lights because it was in a contract that they must be on. It's very frustrating and a great deal of patience and negotiating power is required. There is so much simple stuff to be done, but there is a real drive to do it.

What's the best way to engage people on energy efficiency?

It is no good just putting up a poster, because after a week no one will notice it any more. It's important instead to organise competitions, boat shows, regular e-mails, and set energy champions for every building. Their targets must be regularly reviewed and their progress tracked and rewarded. I advocate a combination of sticks and carrots.

What are you most proud of?

We undertook a project to analyse the consumption across the whole estate of a major high-street retail company and compare the usage per store based on square footage. This exercise highlighted a number of key stores, issues and proposals for us to focus on. Over the next 12 months we utilised our key supply chain and in-house expertise to complete a number of surveys and resultant remedial works.

As a result of these works we achieved a saving in excess of 10 per cent, equating to over £1 million per year. This was accomplished with a budget spend of just £67k. In addition to this, we achieved a collective saving of £350,000 on annual energy consumption through changing to more efficient light bulbs. We are working with the client over the next five years to achieve a further saving of £2 million per annum on energy consumption alone.

6

Passive heating, cooling and air conditioning

Passive heating and cooling refers to techniques to manage the internal temperature and air quality of a building without using power. Consequently, this chapter will examine building design and management with the aim of minimising the use of heating and cooling equipment. It will also look briefly at methods of heat recovery from equipment such as boilers and fridges. Chapter 7 will look at the heating and cooling systems themselves.

In Chapter 4 we introduced the concepts of heat gains, including solar gain, and of the thermal envelope. We looked at different strategies for hot and cool climates. We saw how insulation and airtightness around the building envelope are worthwhile in all climates in order to be able to control the internal climate effectively. Whether a building is being renovated or designed from scratch, it is possible to take advantage of the principles of passive solar architecture to minimise the required artificial energy input in maintaining this internal climate. This typically involves modelling the building design using specialised software.

Chapter 5 examined issues around lighting, including maximising daylighting and optimising windows and glazing. We covered this subject first because it has an impact on heating and cooling. As with the Springfield Literacy Center case study near the end of that chapter, it is advantageous when designing a building to use an architectural design model that also has the functions of energy modelling. The chosen energy modelling software should, likewise, take into account all factors of passive heat gain and loss. By doing so holistically, energy use can be reduced by up to 90 per cent for a new building, or between 50 and 80 per cent for an existing building, depending upon circumstances and budget.

The global warming impact of a building throughout its lifetime can also be reduced by minimising the carbon emissions from the energy used to manufacture the materials for refurbishing and maintaining the building (sometimes called 'embodied energy'), to run and live in the building, and to dispose of the building at the end of its useful life.

Figure 6.1 Thermographic cameras are invaluable for capturing infrared images of buildings to reveal the movement of heat through a building's fabric. Red indicates hot areas, blue cold. For example, in the bottom left-hand picture of a window, it may be seen how cold is coming through from the outside, particularly around the frame. These cameras can be bought, together with software that allows detailed examination of the information they contain. They should be a chief weapon in any serious energy manager's armoury.

Source: Formax 2

The challenge of solar gain

In the daytime, the effective heat storage capacity arising from passive heat gains is governed by the following:

- the thermal capacity of the building;
- the level of insulation;
- how far solar gain penetrates into the building;
- the degree of airtightness.

The available means of dispersing or absorbing this heat include the following.

Day ventilation:

- passive ventilation;
- mechanical ventilation;
- air-to-earth heat exchanger.

Night ventilation:

- passive ventilation;
- mechanical ventilation.

Cooling aids:

- wind towers;
- solar chimneys;
- phase change materials.

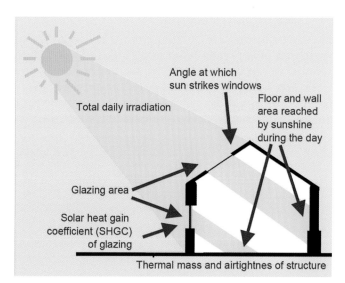

Figure 6.2 Factors affecting solar gain.

Source: Author

Modelling should take account of all of these factors, and their effects throughout the year. For example, the solar gain experienced by the building is a function of the total daily irradiation on the building surface, the glazing area, the angle of incidence at which the sun hits the window, the solar heat gain coefficient (SHGC) of the glazing, and the area of floor or wall reached by the sunlight, as well as its airtightness, U-value and thermal mass.

Options described below for passively managing cooling and heating are available for both new builds and renovations. The applicability of a given

Figure 6.3 Shading used to moderate solar gain on the Federal Press Office in Berlin, Germany.

Source: Frank Jackson

solution for a renovation programme is, of course, dependent upon the particular situation. The ideal situation for medium to large buildings is a building management system that incorporates monitoring for key locations and automatic controls for any of the controllable variables: shading, window opening, venting, heat exchangers. The controls will take account of the way heat moves within the building. This is governed by what is called the stack effect, caused by the fact that heat rises, or moves from areas of high to low pressure.

Calculating the stack effect

The stack effect is a function of the pressure difference between the air outside and inside the building caused by their difference in temperature. This pressure difference (ΔP) creates the stack effect and it can be calculated using the equations below. For buildings with one or two floors, h is the height of the building. For multi-floor, high-rise buildings, h is the distance from the openings at the neutral pressure level (NPL) of the building to either the topmost openings or the lowest openings.

$$\Delta P = C \times a \times h \times [(1/To) - (1/Ti)]$$

where:

SI units:

ΔP = available pressure difference, in Pa
C = 0.0342
a = atmospheric pressure, in Pa
h = height or distance, in m
To = absolute outside temperature, in K
Ti = absolute inside temperature, in K

US units:

ΔP = available pressure difference, in psi
C = 0.0188
a = atmospheric pressure, in psi
h = height or distance, in ft
T_o = absolute outside temperature, in °R
T_i = absolute inside temperature, in °R

The draught flow rate may be calculated using the equation presented below.

$$Q = C \times A \sqrt{[2g \times h \times \{(T_i - T_o)/T_i\}]}$$

where:

SI units:

Q = stack effect draft (draught in British English) flow rate, m²/s
A = flow area of the openings, m²
C = discharge coefficient (usually taken to be from 0.65 to 0.70)
g = gravitational acceleration, 9.81m/s²
h = height or distance, m
Ti = average inside temperature, K
To = outside air temperature, K

US units:

Q = stack effect draft flow rate, ft²/s
A = area, ft²
C = discharge coefficient (usually taken to be from 0.65 to 0.70)
g = gravitational acceleration, 32.17ft/s²
h = height or distance, ft
Ti = average inside temperature, °R
To = outside air temperature, °R

Passive cooling

The stack effect may in theory be used to cool a building without the need for artificial ventilation systems. In practice, with existing buildings it is difficult. The relative size of the openings at the top and ground floor is important. If the local prevailing wind is constant, the best result is obtained by having a small inlet and

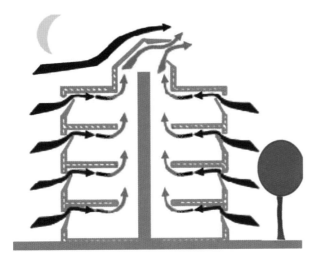

Figure 6.4 The 'passive stack' cooling effect can be exploited by opening selected windows at night. This dramatically cuts down on the need for air conditioning during the day. Window openings near the ground floor would need to be made secure.

Source: Author

a large outlet. If the wind is not constant, a large inlet is preferable because it lets in a greater volume of air.

In the schematic building illustrated in Figure 6.4, the rooms or zones on the outside of such a building, which may be a typical office block, may or may not be separately controlled and air-conditioned, depending upon the degree of sophistication present in the building. If present for most of the time, the air conditioning should need no mechanical input. If hot air is present above a set temperature, it is allowed to escape at whatever rate is necessary to preserve the comfort of this zone's occupants. It escapes into a central vertical space. In a real building this may be a stairwell or lift shaft. It could also be an atrium. At the top of this space, louvres allow a controlled amount of hot air to escape, again at whatever rate is necessary for the comfort of the whole building's occupants. If the building is lucky enough to have an atrium, this is also used to allow solar gain to enter the building.

If available, a courtyard or open space encircled by buildings may be used to increase the passive stack ventilation effect in hot weather. During the day, sunshine coming into the courtyard heats up the air, which rises to escape. To replace it, air is drawn from the building into the space at ground level through openings in rooms facing the courtyard, which in turn must draw warm air out of the building(s).

At night-time, the process is reversed. The warm roof surface is cooled by convection. If the roof is made to slope inward towards the courtyard, the cooler air falls down into it and enters the building. Hot air inside the building is allowed to escape from openings in the top of the building. The larger the thermal mass of the building, the more even the room temperature that will result.

Ground-level vented wall and/or window openings must be equipped with weather, burglar and insect protection and use automatically controlled flaps. If

Figure 6.5 An atrium used to attract solar gain and manage the internal temperature of the Energy Forum building in Berlin, Germany.

Source: Frank Jackson

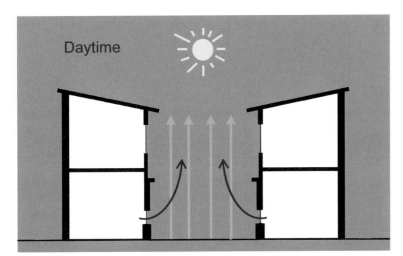

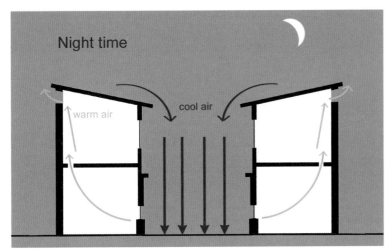

Figure 6.6 (a) Another type of passive cooling: in the daytime, hot air rising in a courtyard draws cooler air from inside the building. (b) During the night, cool air falling from the roof into the courtyard is drawn into the building to replace the warm air rising out of openings at the top.

daytime solar gains are reduced to a minimum and the building is otherwise airtight, night ventilation can perform well. The building must have good passive stack ventilation design.

Air inlets and heat exchangers

Air may be drawn into the building from undergound pipes via vents nearby. In hot seasons or climates, the air will be cooled along the way because the temperature 2 to 3 metres below ground will be cooler than that at the surface. A fan may be required to draw the air through. This may be conceptualised as a primitive heat pump being used as a 'cool pump'. A fully fledged heat pump will have no vent outside, instead consisting of a loop or coil circulating into the ground. (This is discussed further in the next chapter, because it uses energy in order to work. To provide cooling, such a heat pump's heat exchanger would be run in reverse, to transfer the heat from the building into the ground.)

Figure 6.7a and b
An air intake outside for a ventilation system, with a view of the inlet just inside the building (cover removed).

Source: Chris Twinn and Author

Case study: The Energon Building in Ulm, Germany

The Energon passive office building in Ulm, Germany is an example of a new building where such techniques have been employed. It is a triangular, compact building with five storeys, and has a physically curved façade enclosing a glass-covered atrium at the centre. This provides ventilation and daylight. The building is a reinforced concrete skeleton construction with façades made of prefabricated wooden elements of largely equal dimensions. Insulation is 20cm thick under the foundation slab, 35cm in the façade, and up to 50cm in the roof. The windows are thermally insulated triple-glazing. Heat pumps and thermal stores help moderate the temperature.

Figure 6.8
Air intakes for ventilation and cooling outside the Energon passive office building in Ulm, Germany.

Source: International Energy Agency (IEA)

Wind towers

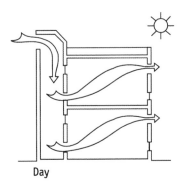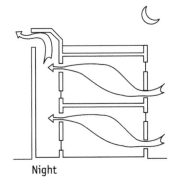

Day   Night

Figure 6.9 Daytime and night-time operations of a wind tower.

Source: K. Bansal, G. Minke and G. Hauser, *Passive Building Design* (Elsevier Science BV, 1994); V.V.N. Kishore (ed.), *Renewable Energy Engineering and Technology: Principles and Practice* (Earthscan, 2009)

Wind towers or chimneys may be seen as an enlarged version of this principle, and are used to encourage the stack effect for cooling purposes. They may be employed in situations where a consistent prevailing wind comes from one particular direction. They may be added on to the side of an existing building. A tower is located on the side of the building to be ventilated. It has openings in the side at the top, designed to face the prevailing wind. Inside, it is separated into two or more shafts, which allow air to move easily up and down the tower at the same time.

In the daytime, ambient air blows into the openings. It is drawn down by pressure differences to the base, and sometimes even underground in a basement, where it cools. It is then allowed to circulate upward through the building and exit through openings near the top. At night-time there is a reversal of airflow; cooler air enters the bottom of the tower after passing through the rooms. It is heated up by the warm surface of the wind tower and leaves in the reverse direction, sucking warm air out of the building. In this way, the heat in the thermal mass of the building, gained during the day, is allowed to escape.

Solar chimneys

Figure 6.10 'Solar chimneys' are automatically opened when required to release unwanted hot air. Their height and metal composition allows them to be heated by the sun, which heats the air internally. This rises through the chimney, drawing up air from within the building. Office building at the Building Research Establishment, England.

Source: BRE

Figure 6.11 Schematic diagram of a solar chimney incorporating a wind tunnel system.

Source: V.V.N. Kishore (ed.), *Renewable Energy Engineering and Technology: Principles and Practice* (Earthscan, 2009)

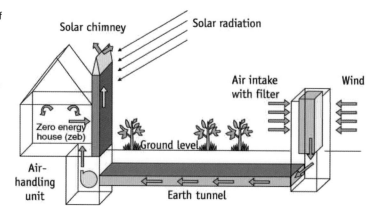

Solar chimneys are employed where the wind cannot be relied upon to power a wind tower. Air within the chimney is heated by solar radiation and rises, sucking out hot air from inside the building. The chimney's outer surface (painted black and glazed) acts as a solar collector to heat the air within it (it must therefore be isolated by a layer of insulation from occupied spaces). Its power depends on the following:

- the size of the collector: the larger it is, the more heat is collected;
- the size of the inlet and outlet: an inverted 'funnel' is best, with the narrow opening at the top;
- the vertical distance between the inlet and outlet: a longer distance creates greater pressure differentials.

Figure 6.12 The roof of the railway station in Berlin, Germany, which uses solar glazing and automatic window openings for cooling.

Source: Frank Jackson

In general, with passive cooling, if free ventilation is not feasible or adequate on its own, mechanical ventilation is added to help it along, but it is useful to remember that it will be more efficient if the mechanical ventilation takes place only at night.

Case study: The Solar XXI building in Lisbon, Portugal

Figure 6.13 The Solar XXI building in Lisbon, Portugal, which functions as a combined office and laboratory at the National Energy and Geology Laboratory (LNEG).

Source: International Energy Agency (IEA)

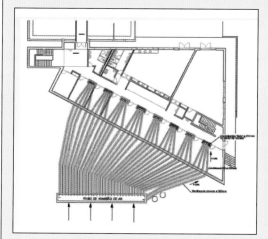

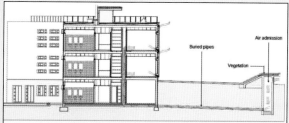

Figure 6.14 and b Plan and sectional view of the Solar XXI building in Lisbon, Portugal, showing distribution of the buried air pre-cooling system.

Source: International Energy Agency (IEA)

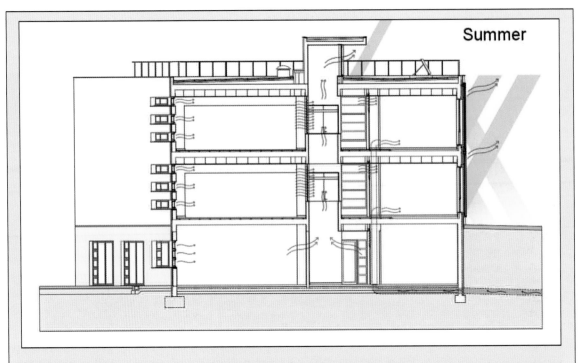

Figure 6.15 Cross- (blue arrows) and vertical (red arrows) ventilation systems acting together with the buried pipes system (blue arrows on the right) in the Solar XXI building in Lisbon, Portugal.

Source: International Energy Agency (IEA)

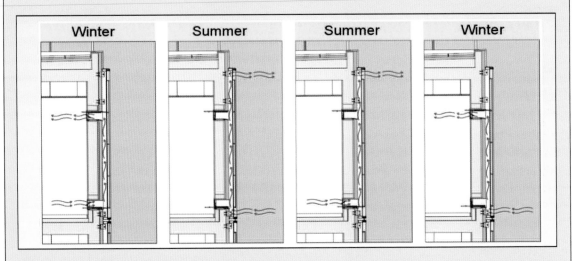

Figure 6.16 The method of operation of the heat output of the PV modules to supplement ventilation (vertical cross-section). There is an air gap behind each panel with openings to indoor and outdoor air at both high and low levels, where heat from the rear of the panel causes a convective flow. In winter, the upper opening takes air indoors, either from outside or from the room, through the lower opening to be heated. In summer, the upper opening lets the warmed air outdoors. The lower opening can either be open to the room to provide ventilation or to outside to provide cooling for the PV panels only.

Source: International Energy Agency (IEA)

This 1500m² (16,146ft²) multipurpose building in Lisbon, Portugal is naturally ventilated and functions as a near zero energy building. Its cost is said to be little more than a conventional building of the same size. The office space is on the south side of the building to take advantage of daylighting and solar heating. Spaces with intermittent use, such as laboratories and meeting rooms, are on the other side of the building. Office spaces are in use from 9 a.m. to 6 p.m. weekdays, and the ventilation pattern has been arranged to suit this pattern.

The building has high thermal capacity and external installation on the walls and roof. The south façade supports 100m² of solar PV modules and the majority of the glazing. Additional space heating is provided by 16m² (172ft²) of roof-mounted solar thermals that also supply hot water, which can be supplemented by a gas boiler. The 18 kilowatt-peak (kWp; the rated power output under standard test conditions) grid-connected PV arrays supply electricity; further panels are located in a car park, where they also provide shade. The entire system satisfies heating requirements of 6.6kWh/m² and cooling requirements of 25kWh/m². Annual electricity use for the building is about 17kWh/m², of which 12kWh/m² is from the PV arrays, leaving 30 per cent to be drawn from the national grid.

Natural lighting is encouraged. In the centre of the building a skylight provides light for corridors and north-facing rooms on all three storeys. The installed artificial lighting load is 8W/m². There is no need for an active (powered) cooling system. Venetian blinds are placed outside the glazing to limit direct solar gain. Natural ventilation is promoted through the use of openings in the façade and between internal spaces, together with clerestory windows at roof level, which help create a cross-wind and stack effect. Assisted ventilation is provided by convection from the heat rising from the PV modules. To supplement this in the cooling season, incoming air may be pre-cooled by being drawn by small fans through an array of underground pipes as shown in figures 6.13 to 6.16. The openings are adjustable, and air is allowed to rise through the central light well. The vents are manually operable, and staff need to be educated in their use. In other buildings these vents can operate automatically, being governed by sensors. The building's occupants have been surveyed and expressed 70 to 95 per cent satisfaction with aspects of the air quality and temperature.

Phase change materials (PCMs)

All substances store energy when their temperature rises and release it when they cool, but when a phase change occurs in a substance (melting or evaporating, condensing or solidifying), the energy stored and released is significantly greater. Furthermore, heat storage and recovery occur isothermally (at a consistent and predictable temperature), which makes them ideal for space heating/cooling applications. Phase change materials (PCMs) may therefore be used to store and release heat to moderate internal temperatures throughout a 24-hour period. Latent heat is the term used to describe the heat released or absorbed by a substance during a change of state or phase.

PCMs use the air temperature difference between night and day, and so running costs are close to zero. In the daytime, incoming external air is cooled by the PCM storage module, which absorbs and stores its heat by changing its phase state (e.g. solid to liquid). At night-time the substance reverts to solid form, releasing its heat by being cooled by the now cooler external air. Commercially available phase change materials come in four categories and are chosen based on the temperature of their phase change relative to that required in the space to be moderated:

1   Eutectics: solutions of salt in water with a phase change temperature below 0°C (+32°F);
2   Salt hydrates: specific salts able to incorporate water, which crystallise during the freezing process, normally above 0°C (+32°F);
3   Organic materials: these tend to be polymers composed primarily of carbon and hydrogen. They mostly change phase above 0°C (+32°F) and can be as simple as coconut oil, waxes or fatty acids;
4   Solid–solid materials: there is no visible change in the appearance of the PCM (other than a slight expansion/contraction) during phase change and therefore no problems associated with handling liquids (e.g. containment, potential leakage, etc.).

For example, an air-conditioning device is available that is integrated with a heat exchanger and uses a phase change material based on expanded natural graphite, for heating or cooling a space. The melting point of this particular PCM package is 20°C (68°F) and its heat capacity is 30Wh/kg. Its rated airflow is around 160m³/h. For cooling loads of up to 50W per square metre of floor area, a rough figure for the mass of the material required is 5.5kg per square metre. Thermal modelling will help size the system correctly.

Plasterboard is available that is integrated with small plastic capsules with a core of wax – a phase change material – inserted during fabrication. The melting temperature of the wax may be defined during manufacturing. If the room temperature rises above this melting point (around 21°C to 26°C (69.8°F to 78.8°F), the wax melts, absorbing as it does so the surplus room heat. Conversely, when the room temperature falls, the wax sets, yielding heat to the internal atmosphere.

PCMs may be used in combination with night ventilation, in this example, to ensure that the wax solidifies at night; otherwise it will not work during the day. The wax capsules are delivered from the manufacturer as liquid dispersion or powder.

Figure 6.17 An office in India whose internal temperature is modified by panels in the ceiling containing a phase change material with a phase change temperature set at 21°C (69.8°F).

Source: PCM Energy, Bangalore, India

Passive heating

Figure 6.18 Allowing sunlight to fall on an exposed tile, concrete, stone or slate floor or wall allows it to absorb the heat which is released later, reducing the need for artificial heating.

Source: Eric Berger, Electric Treehouse

In cooler climates during heating seasons, large areas of glazing on sun-facing walls or roofs are used to allow light to fall on to thermal stores, usually concrete or stone floors and walls, which heat up and keep the heat in the building (see Figure 6.2). On the opposite (North or South Pole-facing) walls, windows should be kept to a minimum and as small as possible for lighting, because heat is more likely to escape through a window than through a wall. Alternatively, they should be well insulated with shutters or curtains. This wall, which receives no sunshine at all, should also have high thermal mass and/or be externally insulated, to retain heat in the building. Large office buildings often employ atria or external sheet glazing as a false 'wall' around the true wall to capture and control heat (see Figures 6.3 and 6.5).

The captured heat is then transferred around the building using the stack effect as described above, or with the help of mechanical ventilation with heat recovery systems (see Chapter 7). Glazing, shading and ventilation are controlled by the building management system to optimise this process so that the occupants do not overheat or become too cold. Some strategies are covered in Chapter 5 under daylighting. This topic is explored in more detail in our companion volume *Solar Technology*.

Heat recovery

Heat may be recovered from many processes that take place within commercial premises, where it would otherwise be lost. The heat may then be used for any purpose, even, perhaps counter-intuitively, for cooling, by driving a heat engine. Low-grade heat can also be concentrated into higher-grade heat. The use of this heat to supplement or even replace fuel used specifically for this purpose can provide rapid returns on investment.

The use of phase change materials, described above, is an example of heat recovery in a basic sense. As we shall see in the next chapter, in a typical ventilation system with heat recovery, the heat from stale air being expelled from a building is transferred, using a heat exchanger, to the incoming fresh air. This is achieved by passing the incoming air over a series of pipes which contain the hot outgoing exhaust air. Efficiencies of these systems can vary from 55 to 85 per cent.

Other sources of 'waste' heat can be: boiler flue gases, boiler blowdown, air compressors, refrigeration plant, hot liquid effluent, power generation, process plant cooling systems, and high-temperature exhaust gas streams from furnaces, ovens, kilns and driers. The most common targets for use of the waste heat are, besides heating incoming fresh air for ventilation: preheating combustion air for boilers, ovens and furnaces, preheating boiler feed water for hot water systems, space heating, drying, preheating for other industrial processes, and space heating. Industrial types of heat reclamation are described in our companion volume *Energy Management in Industry*.

Boilers (UK) or furnaces (USA)

Heat for a building is typically supplied by a boiler, which is called a furnace in the USA. The operational efficiency of the boiler is defined as the useful heat output divided by the fuel input. Many boilers lose heat, for example through the flue (about 18 per cent) and in heat transfer from water or gas. Condensing boilers or furnaces are more efficient than any other type and can achieve around 90 per cent efficiency because latent heat is recovered from the water vapour produced during combustion. Burners should be tuned to the correct temperature by adjusting the fuel-to-air ratio on a regular basis. It should also be checked whether the burner is firing at a rate too high for the boiler to which it is fitted. Together, these measures can improve efficiency by between 5 and 30 per cent.

Boiler flue economisers, if not already present, can be retrofitted to most steam and high-temperature hot water boiler flues and often to non-condensing boilers. Exact savings depend on the type of boiler to which they are fitted:

- If attached to the flue outlet of the most efficient condensing boiler, it reduces the amount of gas used for hot water by an extra 7 per cent;
- If attached to a non-condensing boiler, up to 52 per cent can be saved;
- On an older combi-boiler, up to 37 per cent savings on water-heating expenses are possible.

The heat recaptured may be used to preheat boiler feed water or stored in a tank to provide hot water for other purposes. The economiser must be sized correctly for the flue, and the flue gases must not be condensed to liquid when they arrive at the economiser. Feed water must not boil in the exchanger. If the economiser is designed to condense flue gases, the water returning from the heating circuit must be cool enough to get the benefit of the additional heat, i.e. below 50°C (122°F) for hot water boilers and below 90°C (194°F) for steam boilers.

Preheating the combustion air feeding into the burner to the same temperature as the boiler also improves efficiency, by 1 to 2 per cent. The source for this heat can be the heat remaining in the flue gases, higher temperature air drawn from the top of the boiler house, or heat recovered by drawing air over or through the boiler casing. In the former, outside air is drawn through the boiler flue economiser and ducted to the burner air input. In the second case, air is drawn

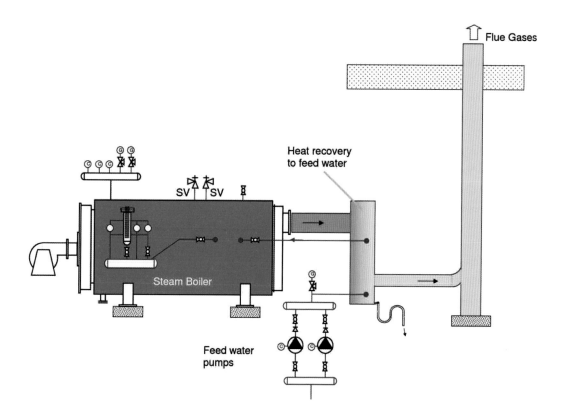

Figure 6.19
A schematic
diagram of a dual
fuel condensing
economiser. Heat from
the waste flue gases
is transferred to the
incoming water.

Source: Public domain
(shared by haj90599)

with a fan from the ceiling level. In a typical office of about 250 people this cheap measure has a rough payback of five years. For a steam boiler, blowdown heat recovery has a similar payback period. The heat is used to heat the feed water. It is unlikely to be cost-effective on boilers below 1000kW.

Heat recovery steam generators (HRSGs)

These are boilers or furnaces that recover exhaust heat from gas turbines in order to generate steam for the steam turbines. Their market is expected to grow substantially, as part of a long-term goal to meet carbon emissions targets by large companies. They are used in combined cycle gas turbine (CCGT) plants.

Refrigeration

Heat recovery from fridges and other equipment is covered in more detail in our companion volume *Energy Management in Industry*. In a typical supermarket, there is a high potential for using the waste heat from refrigeration for space heating, or supplying 75 per cent to 90 per cent of hot water demand.

Having obtained as much 'free' energy as possible, attention may now be turned to managing other energy inputs for heating and cooling, which is discussed, along with degree days, in the next chapter.

Energy managers share their experience

James White, Senior Energy Conservation Engineer at Chelan County Public Utility District, Washington State, USA

What does your job involve?

I work for a publicly owned electric utility and run their commercial and industrial energy efficiency programme. I work mainly with the industrial fruit warehouses (apples, pears and cherries), but also help other businesses save energy. I design the conservation programmes that we offer and then distribute financial incentives to help these customers conserve energy. The electric utility I work with generates all of its energy from two large hydropower projects that were built over 50 years ago. As a result of those earlier investments, we have some of the least expensive power in the nation. It is also carbon emission-free. We also do a good job of protecting the salmon and other fish that might be impacted by our dams on the Columbia River.

What qualifications did you gain?

I have a Bachelor of Science in Mechanical Engineering degree from the University of Alaska in Fairbanks and a Masters and Doctorate degree specialising in energy management from Texas A&M University.

What does your day-to-day work look like?

I talk to customers about conservation opportunities at their facility, calculate the expected energy savings, complete the paperwork and process their payment requests when the come in. I also design new incentive programmes, such as the irrigation pump programme that can help farmers irrigate their crops, while using less energy and giving them better control over the water pressure. To confirm that we are actually achieving the energy savings we predicted, I also install recording meters to measure the actual energy savings.

What do you love about your job?

I love having the freedom to work independently within a quality company providing a win-win-win-win for everyone. The energy we conserve frees up additional hydropower that reduces the amount of fossil fuels generated some-where else. The savings also reduce our customers utility costs. We can make more money selling power on the wolesale market than we do selling it to our customers, so conservation provides positive cash flow to our utility and our customer/owners.

What do you think are the biggest challenges?

Living in a community where a majority of the population is very conservative (65%) and ignorant of the impact mankind is having on the planet. Ignorance is not lack of knowledge, it is the deliberate act of choosing not to know some-thing.

What are you most proud of?

I am most proud of the green power programme I came up with to make solar and small wind projects cost-effective in our area, even though we have some of the lowest electric rates in the country. The programme was also adopted by two electric utilities in Alaska (Golden Valley Electric and Homer Electric Company). Because of the Sustainable Natural Alternative Power (SNAP) programme that I came up with, there are many solar and wind projects that exist that otherwise would never have been built. I am also working with the Electric Power Research Institute to develop a simpler/easier way to meter solar power generation that I believe is going to transform solar power by bringing electric utilities into the game. The idea is to have electric utilities take over ownership of solar generated power at the DC voltage. Electric utilities would meter customers' solar generaton at the DC side of the inverter and treat the inverter just like we do standard AC-to-AC transformers. The electric utilities then become responsible for all the grid interconnection details and hassles. The utilities also get full cost recovery and the customers only need to deal with the solar modules, which is the longest lasting, most reliable part of the whole system.

What's the best way to engage people on energy efficiency?

Most people get it. The projects drive themselves because they usually do more than just reduce energy costs. For example, they usually get better light, improved operations and lower maintenance.

What is the first thing that you would recommend to save energy and carbon emissions?

Change people's perception about energy. We were told we could never reach the Kyoto limits, but massive hydraulic fracturing of horizontal wells has created an abundance of natural gas that is displacing king coal and driving us down to 1990 levels of CO_2 emissions. Getting people to envision a fully renewable future where we are able to do many more things while using far less energy than we do now.

What is your favourite tool of your trade?

The education I received and Excel spreadsheets.

7

Active heating, cooling and air conditioning

In the last chapter we examined the use of various strategies to minimise the need for specific heating and cooling equipment. But sometimes it is necessary to use additional energy for heating, ventilation and air conditioning. This chapter investigates strategies for doing so in as efficient and climate-friendly a manner as possible, including the various fuel and technology choices available.

Heating and cooling systems are generally considered together because they combine throughout the year in any climate to moderate the internal atmosphere and temperature. Their joint aim is, for the least amount of energy input, to maximise comfort and health while controlling humidity and air quality. In any poorly managed building one can generally find examples of equipment increasing the temperature inside a building, while a cooling system is struggling to reduce it. Therefore the two should be managed together, and within the context of the building's design and its energy management system (BEMS). To determine how much energy is required for heating or cooling a building, and whether the required amount is actually being used, it is useful to employ degree days.

Degree days

A heating degree day is worked out relative to a base temperature building occupants can tolerate without needing the heating on; in the UK the predominant convention is to use 15.5°C (59.9°F). Elsewhere it varies, with anything between 60 and 65°F (15 to 18°C) being used in the United States. Once chosen, however, the baseline must be consistently used for meaningful results. Factors in the choice of baseline temperature are the other sources of heating within a building, such as body heat and heat given off by equipment, which can raise the internal temperature to normal comfort levels, generally taken to be between 18 and 20°C (65 to 68°F).

To find out the heating degree day for a given day, take the average temperature on any given day and subtract it from the base temperature. If the result is above zero, that is the number of heating degree days on that day. If the result is below or equal to zero, it is ignored.

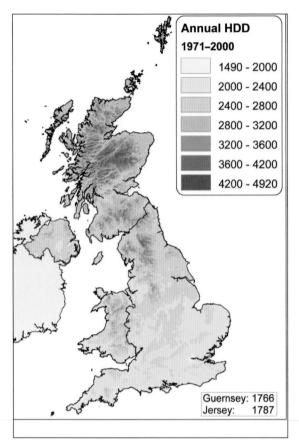

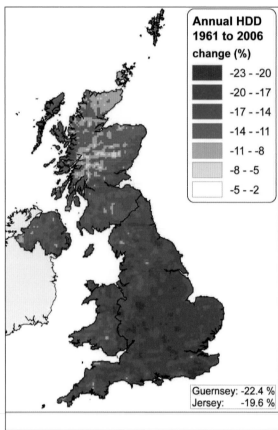

Figure 7.1 Annual average heating degree days in Great Britain and Northern Ireland for 1971 to 2000.

Source: Crown copyright/Met Office, UK

Figure 7.2 Percentage change in annual heating degree days in Great Britain and Northern Ireland from 1961 to 2006. All areas show a reduction, but particularly the south, which could be said to reveal the regional effects of climate change.

Source: Crown copyright/Met Office, UK

Figure 7.3 Graph showing annual heating degree days from 1961 to 2006 filtered by UK region, based on a linear trend. The effect of climate change can be clearly seen in the falling lines. Less energy may be being used to heat buildings, but more is being used to cool them.

Source: Crown copyright/Met Office, UK

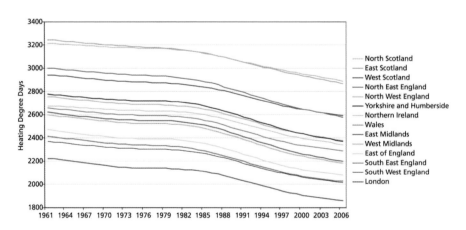

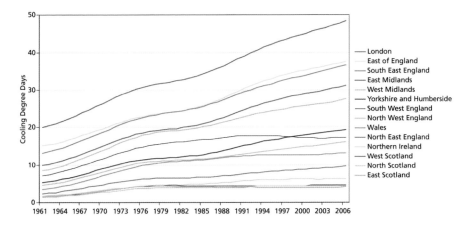

Figure 7.4 Graph showing annual cooling degree days from 1961 to 2006 filtered by UK region, based on a linear trend. The effect of climate change could be said to be revealed by the rising lines. The result is that more energy is being used to cool buildings, resulting in more greenhouse gas emissions. This vicious circle can be broken using passive cooling, and low-carbon cooling techniques, outlined in this chapter.

Source: Crown copyright/Met Office, UK

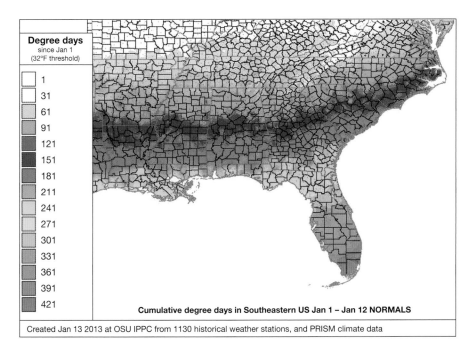

Cumulative degree days in Southeastern US Jan 1 – Jan 12 NORMALS

Created Jan 13 2013 at OSU IPPC from 1130 historical weather stations, and PRISM climate data

Figure 7.5 Sample cumulative degree days map for southeastern USA, 1–12 January.

Source: GIS GRASS 5.4

For example, a location on one day might have a maximum temperature of 14°C (57°F) and a minimum of 5°C (41°F), giving an average of 8.5°C (47.3°F). Subtracted from 15.5°C this gives 7°C (44.6°F). A month of 30 similar days might accumulate $7 \times 30 = 210$ degree days. A year (including summer temperatures above 15.5°C) might add up to 2000 degree days.

The rate at which heat needs to be provided to this hypothetical building corresponds to the rate at which it is being lost to the outside. This rate, for a 1° temperature difference, is simply the U-value (R-value in the USA and Canada) of the dwelling (as calculated by averaging the sum of the U-values of all elements) multiplied by the area of the dwelling's external surface.

The U-value of a material is a measure of its ability to resist heat loss. The lower the U-value a material has, the better insulation it provides. In the USA, R-values are used instead of U-values. They are the inverse of U-values. Therefore, the higher the R-value a material has, the better insulation it provides. To convert R-values to U-values, divide into 1. Again, the R-value of the dwelling is calculated by averaging the sum of the R-values of all elements.

Multiplying the rate at which a building is losing heat by the time (in hours) over which it is losing heat reveals the amount of heat lost in watt-hours (Wh) or Btu: this is exactly the amount of heat that needs to be provided by the heating sources. (To convert Wh to kWh, divide by 1000.) Therefore, if the external area of the dwelling (walls, roof and floor) is A, its average U-value is U and the number of degree days is D, then the amount of heat required in kWh to cover the period in question is (for Europe):

$$A \times U \times D \times 24/1000$$

where the multiplier 24 is needed to get the value in kWh rather than kWdays or simply Btus in the case of US units. The same method is used (without dividing by 1000) for US units with the result being in Btus rather than in kWh.

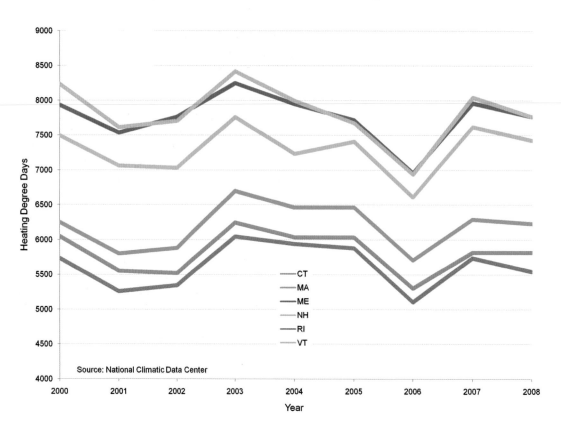

Figure 7.6 Heating degree days for selected states in the USA, 2000 to 2008.

Source: US National Climate Data Center

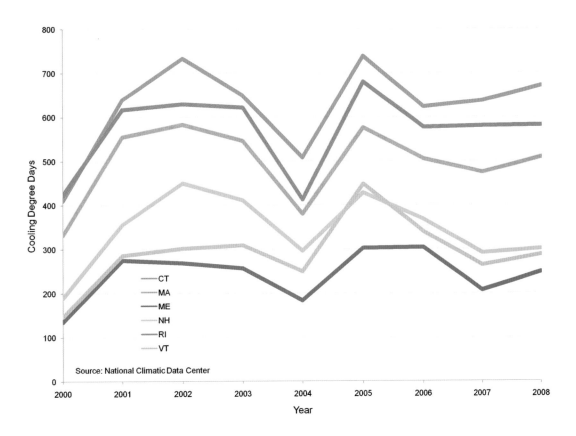

Figure 7.7 Cooling degree days for selected states in the USA, 2000 to 2008. The average trend is upward in all cases, most likely indicating the influence of climate change.

Source: US National Climate Data Center

For example, if the building surface area is 500m², the average U-value is 3.5 and the degree days number 2000 then the annual amount of heat required is:

$$500 \times 3.5 \times 2000 \times 24/1000 = 84,000\text{kWh}$$

The above is a simplified case. The reality is more complicated. There are additional factors to take into account, such as the thermal properties of the building, wind chill, exterior shading, heat gains within the building, the heating schedule and solar gains. In addition, heat requirements are not linear with temperature, and well-insulated buildings can have a lower target point.

Nowadays, software is available to help with these calculations, but the appropriate figures still need to be entered. If reliable historical records of the building's energy consumption are available, the base temperature can be worked out experimentally, using linear regression analysis (see p. 131). Degree days are available worldwide for cooling and heating calculations at www.degreedays.net, and have been freely available since 1970, thus providing a sound base for making reliable calculations. Calculations may also be aided by using the website www.wolframalpha.com and typing in 'degree days'.

Figure 7.8 The brown area under the red baseline represents the degree day value for the period, or an indication of the amount of heat that needs to be supplied to make up the difference between the base temperature and the free energy from the sun, represented by the blue line.

Source: Author

Using degree days to map heating or cooling requirements

The degree day value

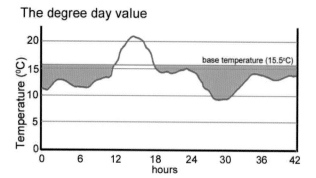

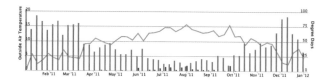

Figure 7.9 Temperature and degree days in Edinburgh, Scotland, plotted over a year (2011). The green line represents hourly temperatures averaged over time, red columns represent heating degree days, and blue columns represent cooling degree days (both calculated per day and visualised as period averages).

Source: Author

A graph of heating degree day values plotted against weeks, superimposed over a graph of fuel for heating during the same period, should show a strong degree of correspondence. For the energy manager, the challenge is to calculate whether there is an improvement over time, in order to convince senior management to invest in insulation and draughtproofing. This may best be shown by an example. It is useful to compare energy use during the same month over two consecutive years to which intervening time insulation has been added. One would expect energy consumption for heating in year 2 to be less than in year 1. In fact, the figures are as follows:

Heating energy consumption in month 6, year 1: 34,376 kWh
Heating energy consumption in month 6, year 2: 35,753 kWh

At first sight, it would seem that the insulation strategy has failed. In fact, when weather data are examined, the energy manager notices that the weather in year 2 for that month was colder than year one. Looking up the degree days, the following is found:

Heating degree days in month 6, year 1: 296
Heating degree days in month 6, year 2: 335

Dividing the energy used by the degree days, the following is arrived at:

kWh per degree day in month 6, year 1: 34,376/296 = 116.14
kWh per degree day in month 6, year 2: 35,753/342 = 106.72

In other words, less energy was required to heat the building in year 2 for the same amount of heating needed. By dividing the first figure into the second figure, we can actually see in percentage terms the savings achieved: 90 per cent of the energy used in the first year was used in the second year, representing a saving of 10 per cent.

To calculate cooling degree days, simply apply the reverse of the above system. In other words, the degree days are noted according to the number of degrees the average outside temperature is each day above a baseline temperature that is considered comfortable, which is usually the internal temperature at which the ventilation and cooling system is switched on. Degree days are then added up when the temperature outside exceeds this baseline temperature. The above calculation can then be repeated if the energy requirement for cooling is known.

Degree days and temperature monitoring

The above system can be made even more sophisticated if outside temperature monitoring data are available. Matching hourly or half-hourly temperature readings with variations in heating demand will provide a much more accurate pattern. After all, temperature outside can vary constantly.

There are other ways of working out how much heating or cooling capacity is required, for example, the peak load requirement. More information, together with free data, may be found at www.degreedays.net.

Linear regression analysis

Linear regression analysis is a way of correlating energy consumption data with degree day data. The sources may be detailed interval data from a smart meter, or weekly records of energy consumption. Most buildings follow a weekly pattern of energy use, making this period of time ideal for regression analysis. Monthly analysis will not be useful, since months do not contain a whole number of weeks, making comparison between months difficult.

The longer the time period over which data can be gathered the better: several years is perfect, but a full year will do. As above, separate calculations will be needed according to whether it is the energy requirement for heating or for cooling the building that is being investigated.

The data are entered into a spreadsheet (or specialist software). Column 1 represents the weeks, column 2 the heating degree days, column 3 the energy used in kilowatt-hours or Btus, or volumes of fuel. The second and third columns of data are used to plot a scatter chart. This will show heating degree days against energy consumption.

If Excel is being used, right-clicking one of the data points will reveal the option to select 'add trendline'. Selecting 'linear' and 'display equation on chart' under 'options' will enable the display of a straight line, representing the average fuel requirement, and the graph should display an equation with an R^2 value. In statistics, R^2 is known as the coefficient of determination. Its main purpose is the prediction of future outcomes on the basis of other related information, and is usually seen as a number between 0 and 1.0, that describes how well a regression line fits a set of data. Here, this value will represent the gradient of the trendline and the intercept, which is the point at which the trendline crosses the Y axis.

Example

Month	HDD	kWh
January	355	1392
February	323	1192
March	245	978
April	121	476
May	98	308
June	64	206
July	22	198
August	9	205
September	55	285
October	104	357
November	200	592
December	289	832

The intercept is a significant point because it represents the baseload energy consumption, or the amount of energy being used that is not directly involved with heating or cooling the building, such as for equipment, industrial processes and lighting. In our example it may be read from the graph at about 90kWh. Degree day calculations assume, rightly or wrongly, that the baseload energy consumption is constant throughout the year. This use of the word baseload is not the same as that used elsewhere. 'Baseload' is always relative to what is being calculated. If we are calculating the amount of energy used in a restaurant, the baseload would represent all the energy in the building that is used in cooking-related processes. Note that this energy will still give off heat.

Figure 7.10 and the accompanying equation make it possible to predict energy consumption for any degree day by simply reading it off the scale. It then becomes possible to compare predicted energy consumption with actual energy consumption at a given point in time. This enables the energy manager to see whether predicted energy savings have been made following the subsequent introduction of energy-saving actions.

The R^2 value which the Excel spreadsheet has generated provides an indication of how accurate the correlation is. The closer the R^2 value is to 1, the better the correlation, and the more efficiently the heating or cooling system is working. A correlation of 0.9 or above would be very good, whereas one below 0.7 would show either that controls were very poor or that the methodology used to obtain the result contained errors.

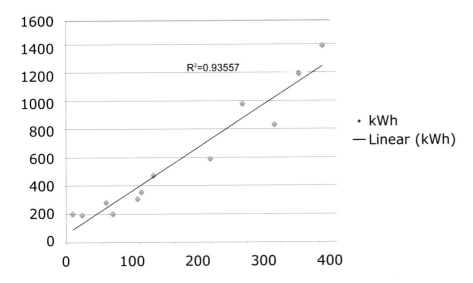

Figure 7.10 Using Excel to analyse linear regression and obtain a R^2 value. The graph is obtained from the columns in the table above. See text for explanation. Along the bottom on the X-axis are kilowatt-hours and up the side on the Y-axis are heating degree days.

Source: Author

The base temperature of the degree days chosen has a huge influence on this correlation factor. This is because the optimal base temperature will vary from building to building, and depend on the heat given off by equipment. The R^2 figure therefore gives a good idea of whether the base temperature chosen is the correct one.

CUSUM

Cumulative sum charts (CUSUM) are used to calculate deviations from expected energy usage, taking into account degree day data. The cumulative sum of daily, weekly or monthly energy consumption is plotted against time. This gives an upward slope when graphed. It is then adjusted for degree days. Finally, it is compared to the previous year by subtracting one from the other for each time period. If savings have been made, the numbers in the second year will be lower than the corresponding figures in the first year. This will be represented in the resulting CUSUM curve going down. If it is rising, the contrary is the case. The curve is created by plotting column 6 against column 1 below. If this process is to be executed by employing a spreadsheet, then the columns are, from left to right:

Month/week/ day	Degree days (kWh)	Actual consumption (kWh)	Predicted consumption (kWh)	Difference (kWh)	CUSUM
	From published or calculated values	From meters	Gradient of the trendline (from linear reduction graph) × column 2 minus the baseload (intercept point on Y-axis)	Column 3 minus column 4	Cumulative sum of figures from column 5

Conventional building

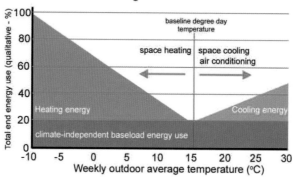

Energy-efficient building

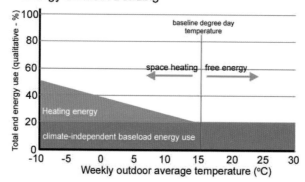

Figure 7.11a and b The graphs indicate the percentage of energy use in a conventional building (a) and an energy-efficient building (b) in a temperate climate. The green rectangle at the base represents the baseload energy use for other purposes. The vertical red line represents the base temperature chosen for degree day calculation (15.5°C/59.9°F). To the left of this line energy is required for heating, and to the right, for cooling. In the energy-efficient building, no energy is required for cooling purposes. When it is very cold, 80 per cent of the non-baseload energy consumed in the building is for heating, whereas comparatively less, 30 per cent, is used in the energy-efficient building. The height of the blue triangles indicates the proportion of energy needed for heating at different temperatures below the base temperature.

Source: Author

A detailed worked example may be found in the free booklet *Degree days for energy management – a practical introduction* (CTG004 Technology Guide), downloadable from the Carbon Trust website at www.carbontrust.co.uk. A similar curve may be created by subtracting actual from predicted figures for the same time period. Where the actual figures are lower than the predicted figures, the curve will go down. Where energy consumption was higher than predicted, the curve will rise. A CUSUM usually has both upward and downward sloping parts. If they are predominantly upward then there is a tendency to use more energy than expected. The converse shows that performance was better than expected.

These calculations, based as they are on real data and real weather conditions, are used to plan energy purchasing budgets, to evaluate and monitor the success of energy efficiency measures, and to help plan further measures.

Sources of heating and cooling

In order to minimise running costs and greenhouse gas emissions, heating and cooling requirements should be supplied, as much as possible, from passive sources or reclaimed heat, as described in the previous chapters. To provide for any shortfall in supply, what, then, are the most efficient and low carbon solutions? The remainder of this chapter will look first at individual technologies and then at integrated HVAC systems.

Heating can be supplied by electricity, gas, solar thermal, heat pumps (air, ground and water sourced), oil and biomass. Electricity may be locally produced and renewable; or grid-sourced. Oil may be fossil fuel based or sourced more sustainably from biofuels and recycled oil or cooking oil. Equally, gas may come from the gas grid, be LPG (liquid petroleum gas), or it could be renewably produced biomethane from anaerobic digestion plants or landfill gas.

In terms of overall climate change impacts, the favoured options would be as follows:

1   Sustainably sourced biomass (but see page 185), gas from renewable sources (biomethane from anaerobic digestion), and oil from renewable sources (biodiesel) or recycled oil. All of these would feed combined heat and power cogeneration (CHP), meaning that the same plant would produce both heat and electricity, the most efficient type of heating plant (up to 95 per cent efficient). Most successful plants of this nature feed the hot water through a heat main to nearby buildings; a district heating system. The electricity produced may be used or sold locally, or exported to the grid.

2   Directly supplied, locally produced electricity, whether from CHP or renewable sources such as hydroelectricity, wind or photovoltaic solar, is more efficient than grid electricity because transmission costs over the grid may lose up to 10 per cent of the original power and result in higher supply costs.

3   A solution involving a combination of heat pump and/or solar thermal is also an option, with the heat output being stored in water tanks. It is even more desirable if the electricity powering the pumps is renewably sourced.

4   The next favoured options would be: energy-from-waste CHP, then CHP using gas, either LPG or from the grid. Gas has half the carbon impact of coal

Table 7.1 Classification of heating fuels

Gas	Oil	Biomass	Solar	Heat pumps	Electricity	
Renewable	anaerobic digestion	used vegetable oil	wood pellets	Solar thermal panels (flat plate collectors)	ground source	wind
	landfill gas	used mineral oil	logs	evacuated tubes	air source	solar photovoltaic
		biodiesel*	woodchips	concentrated solar thermal	water source	hydroelectric
			waste/garbage		geothermal	fuel cell
Non-renewable	LPG	mineral oil	coal			coal
	gas mains					gas

Note * Biodiesel may be produced from soybeans, canola, oil palm or sunflowers (not so sustainable), animal fats and used cooking oil (the most sustainable). Some is produced from commercially available methanol, which, is often made from the methane that exudes from oil and gas wells; burning it, or a methyl ester derived from it, continues to exacerbate climate change.

and oil (around $400kgCO_2/MWh$). For smaller buildings a condensing boiler is an efficient option.

5   In general, using electricity for heating is not recommended, if coming from the grid, in areas where the grid has a high proportion of fossil fuel-burning power stations. This is because it is over three times more efficient and less polluting to burn the fuel directly for heating space or water than it is to convert it to electricity first, transport it, and then turn it back to heating.

6   Finally, to be avoided if at all possible, are oil- or coal-fired heating, because of their impact on climate change.

Renewable electricity is discussed in Chapter 9. The remainder of this chapter discusses renewable heat.

Condensing boilers (furnaces)

Condensing boilers (furnaces) will be the choice for many buildings. These use a hot water tank for storage and contain a combustion chamber through which pass the hot gases created by burning gas or oil. This is surrounded by a heat exchanger which transfers heat from the gases into the water inside. The cooler, heavier flue gases are blown out of the boiler by a fan. Next, a second heat exchanger removes even more of the heat from the flue gases. It pre-warms water coming back into the boiler from a heating system, so less gas is required to heat the water. By using this otherwise wasted heat, condensing boilers use around 90 per cent of the heat they generate.

Since the waste gas loses some of its heat, it cools down into an acidic water called condensate and water vapour, giving the boiler its name. The cooler,

Figure 7.12
A Viessmann Vertomat condensing boiler installed in a multi-unit apartment building in White Rock, British Columbia, Canada.

Source: Wikimedia Commons (Dunnd74)

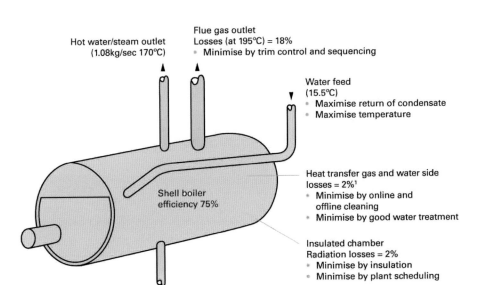

Hot water/steam outlet
(1.08kg/sec 170°C)

Flue gas outlet
Losses (at 195°C) = 18%
• Minimise by trim control and sequencing

Water feed
(15.5°C)
• Maximise return of condensate
• Maximise temperature

Heat transfer gas and water side losses = 2%[1]
• Minimise by online and offline cleaning
• Minimise by good water treatment

Shell boiler
efficiency 75%

Insulated chamber
Radiation losses = 2%
• Minimise by insulation
• Minimise by plant scheduling

Water outlet
Blowdown losses = 3%
• Minimise by good water treatment
• Minimise blowdown heat recovery

Figure 7.13 The sources of losses from shell boilers (the most common type). Most boilers of this type can be made to run at over 80 per cent efficiency.

Source: The Carbon Trust

heavier flue gases are blown out of the boiler by a fan. Manufacturers claim that up to 98 per cent thermal efficiency can be achieved; but a field trial conducted by the Energy Saving Trust[1] in the UK found an average efficiency of 85.3 per cent. Because the water they heat is stored in a tank, they may be used in combination with other heat sources such as a heat pump, solar water panels or electric immersion. Always choose an A-rated boiler.

If the building has a non-condensing boiler, it is possible to upgrade it relatively cheaply by fitting a passive flue gas heat recovery system to the flue above the existing boiler to capture some of the lost energy and use it to heat the water. Fifteen to 30 per cent of the total heat can be reclaimed this way.

Combi boilers (furnaces)

Combination or combi boilers do not use a tank, instead supplying hot water on demand. They combine the central heating with (tankless) hot water supplies in one unit. In the same EST trial, these recorded an average efficiency of 81.5 per cent. Combi boilers can produce between 9 and 18 litres of water a minute (when the cold water is heated up by 35°C/95°F). They will generally provide high water pressure to just one tap at a time. They are not suitable for buildings where more than one use for hot water might be required simultaneously. There is also a delay between turning on the tap and the water heating up.

Both condensing and combi boilers should be sized on the basis of the hot water requirements. There is a trend towards increasing the electricity use of these boilers by fans, pumps and control systems. Average combination boilers use around 30 per cent more electricity to supply 10,000kWh of heat to regular boilers, and around 50 per cent more to supply 20,000kWh of heat. This consideration should be a factor in the choice of heating supply.

Larger boilers can have their efficiency improved in three ways:

1   Improving their combustion efficiency, by checking the operation of the burner, its controls, and removing unburnt fuel, soot and ash;
2   Improving transfer efficiency, by removing deposits on the heat transfer surfaces, using specialised water treatment;
3   Minimising boiler heat loss, through heat reclamation from the flue gas (see p. 136) and by insulating the boiler, including checking for any damage to the insulation and repairing it.

CHP or cogeneration

Combined heat and power (CHP), or cogeneration, removes the need to have separate electricity generators and boilers. A cogeneration plant is a generator that uses heat recovery to reclaim most of the 60 per cent or so of the energy output that is otherwise lost as heat. Usually, a site wishing to install such a system will commission an external supplier who can offer a professional partnership arrangement for developing new energy schemes, and can provide all or part of a system. Typically, this is known as an Energy Services Company (ESCo). The power purchase agreement covers delivery of the power and heat, which means that the plant is delivered as a turnkey operation, and the ESCo is responsible

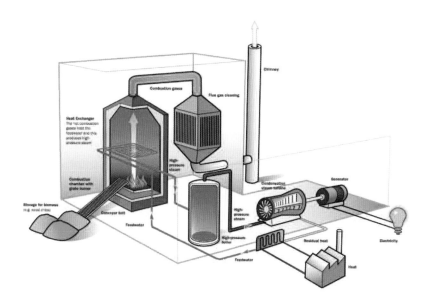

Figure 7.14
Schematic diagram of how a combined heat and power, or cogeneration, plant works. In this case, the plant runs on biomass (woodchips or pellets).

Source: Wiki Commons: Jonathan O'Reilly

for all maintenance. An ESCo may also supply some of the larger other types of heating equipment described below.

CHP plants may also be used to supply cooling. Then it is known as tri-generation or CCHP. The waste heat is captured and passed through an absorption chiller. The electrical power from the unit will help power the pumps that circulate the chilled water. This is ideal for sites that require a large amount

Figure 7.15 A series of mini-CHP units. These have a small commercial-scale application and are rated with an electrical output of 4–13kW and a thermal output of 17–29kW. Units sized much smaller than this are not usually economical to run unless they are switched on constantly.

Source: EC Power

of air conditioning. It displaces the need for separate air conditioning, reducing overall carbon dioxide emissions by as much as 29 per cent. Substantial savings are possible when installing such a system from scratch.

While large CHP plants are the province of industrial sites or local authority schemes, mini-CHP systems are available to meet the requirements of large and medium-sized stand-alone buildings such as offices, hospitals, leisure centres, hotels, supermarkets, care homes and so on. For example, supermarket chain Sainsbury's has two stand-alone packaged gas-fired CHP units at a store in West London, each sized at 210kW. These have reduced energy bills by £20,000 per year, and CO_2 emissions by almost 2000 tonnes over the same period.

Micro-CHP systems are even smaller. The size of domestic fridges, they run exclusively on gas and produce up to 1kW of electricity per hour. Units need a consistent heat load to supply for a minimum of 11 hours per day throughout the year, for 17 hours per day for two-thirds of the year in order to be economic. Heat and electrical requirements must coincide, and the electrical load must exceed supply of the plant. Mini- and Micro-CHP systems must be sized to match the base heating load so that they operate not intermittently but for many hours at a time, making the value of electricity generated pay for the marginal investment in as little as three years. The property's needs for heat and electricity must coincide throughout the year, and the electrical load in the property must exceed the supply of the plant. It therefore works best with a buffer storage tank to save the surplus heat for later, and with a grid connection for electricity export.

Biomass boilers

Biomass is assumed to be carbon neutral, because the vegetation burnt is replaced by a new plantation which, in turn, will recapture the carbon dioxide from the

Figure 7.16 This biomass woodchip boiler is fed from a large hopper, or storage container, above ground. Note the well-insulated pipes.

Source: Author

atmosphere. However it is not entirely carbon neutral, depending on how far away the fuel is sourced, and whether it is in the form of pellets – which are energy intensive to produce – basic timber or woodchips. Therefore the fuel should be reliable, locally sourced, accessible and appropriate. As an example, a 150kW woodchip boiler will use around 400 tonnes of seasoned woodchips a year. This would require around 5000 to 7000 acres of nearby woodland from which the pollarded or coppiced wood could be taken under a service agreement to do so. The fuel's moisture content also affects is calorific value, so it must be dried and seasoned.

Case study: Haulbowline Naval Base, Ireland

A waste oil boiler was incorporated into a building upgrade for mixed-use workshops, offices, toilets and canteen at Haulbowline Naval Base, Ireland. M&E consultants found that every year naval ships produce 70,000 litres of usable waste engine oil, while ships' kitchens and cookhouses produce 12,500 litres of waste vegetable oil. They were instructed to install a boiler fired by this waste oil to provide space heating and hot water to the facility.

The volume of oil required to fuel the boiler was estimated to be 25,000 to 35,000 litres per annum. If diesel were used just 20,000 litres would be required, due to the higher energy content in diesel compared to waste oil. The additional capital cost of the boiler was €7300. Design and consultancy fees were €2500. Additional valves, etc. cost €1800. The total extra cost for installation of the waste oil boiler was €13,166 including 13.5 per cent VAT. This saved 20,000 litres of diesel at €0.6 per litre, or €12,000 per annum. The payback period was 13 months.

The yearly additional running cost due to filters, extra maintenance and so on was €1400 including VAT, so after 13 months there have been savings of €10,600 per annum. As oil prices increase, savings increase. If used as the only oil source, vegetable oil would make the boiler system more low carbon. Local restaurants may provide such oil free of charge. This project has potential for expansion to other sites in the defence forces.

Figure 7.17 Haulbowline Naval Base, Ireland.

Source: Irish Department of Energy

Every attempt should be made to size the plant correctly, since if it is over-specified, which is often found to be the case, it will be running inefficiently if it is required to output at a lower temperature than its optimum operating temperature. Biomass boilers operate automatically to a certain extent but they do require more supervision than a gas boiler. They may be auto-fed from multiple twin containers fitted with a rotary arm fuel extractor and side-opening doors, which allow the fuel to be chipped and blown directly into the fuel store. If wood pellets are used as a fuel, these may be loaded into a hopper and auto-fed into the boiler, and again left alone for a period of days.

Reused oil as a fuel

The case study below from the Irish Navy is an example of the use of recycled oil as a fuel for heating. This is ideal if an organisation produces a large and reliable supply of used oil which would otherwise be thrown away. There may also be a local source of used vegetable oil, for example, from catering businesses, or of biodiesel, which is chemically almost identical. All of these may be burnt in a CHP plant. Existing oil-burning boilers may be converted to consume such fuels.

Anaerobic digestion (AD)

Anaerobic digestion (AD) provides biogas that may be used to supply heat. AD is a type of accelerated composting in which the feedstock may be any organic material: food, and agricultural and gardening waste. Anaerobic digestion has multiple benefits: as well as providing biomethane, it also produces a fertiliser which may be applied directly on the ground or sold. A guaranteed year-round supply of organic waste material to use as a feedstock is necessary. Therefore, the owner must either have this available themselves (they may be a food producer or retailer, or an agricultural concern) or source additional material from similar nearby companies, who may pay to have their material taken away, since they would normally pay disposal costs.

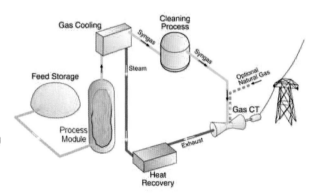

Figure 7.18 Schematic diagram of an anaerobic digestion plant. The composting of the organic material (feedstock) occurs in the process cylinder.

Source: Author

The biomethane may either be used directly to produce heat and electricity in an on-site CHP plant, or even used as a fuel in some vehicles. Surplus gas may also be sold to the gas grid after being treated, and the electricity from the CHP plant may be exported or used locally.

Energy-from-waste

The incineration of waste is less efficient than anaerobic digestion, and should therefore be reserved for non-organic materials. It entails burning the dried waste to drive a steam-powered CHP plant. Like AD, energy-from-waste requires a constant input of suitable combustible material. Additional challenges involve drying the waste and ensuring that no pollutants are released into the atmosphere, in order to meet the requirements of local pollution regulations and the environmental monitoring regime.

Opponents of this technology argue that it discourages the recycling and reuse of waste, and that even legislation-compliant plants produce some air pollution. An incinerator may be useful if a facility produces a constant amount of a particular waste that cannot be disposed of in more sustainable ways and would otherwise go to landfill; at least the energy content of the waste material will be recovered.

Heat pumps

Heat pumps use the free heat from the sun that is stored in the ground or the air, and concentrate it in a much smaller area (i.e. a building) using a pump. Heat pumps are an efficient technology because the ratio of the energy input in the form of electricity to power the pump to the energy output in the form of heat is designed to be at least 3:1. This ratio is called the coefficient of performance (COP), and the higher it is, the more efficient the unit. In this case, for each unit of electricity put in, three or more units are produced.

The exact efficiency found in practice depends upon the difference in temperature between the target and the source at any one time. If the target temperature is underfloor heating at 18°C (64°F) and the source is 2°C (35°F), less energy is required to concentrate and pump the heat than if the source is colder (as in air-source heat pumps in freezing weather) or if the target is hotter (as in a radiator-based heating system where the target temperature may be 60°C/140°F). This is why ground- and water-source heat pumps are generally more efficient than air source; in winter, the temperature of the air outside is generally much lower than that of the ground three metres below the surface.

A heat pump operates in a similar way to the refrigerator, i.e. it includes a refrigerant that passes through a compressor and a condenser. Heat pumps can be confused with geothermal energy. Strictly speaking, geothermal energy is derived from hot rocks that can be relatively close to the Earth's surface in some parts of the world. It requires a very deep borehole to access them: 0.5 to 1 kilometres. Water is pumped down, heated up by the hot rocks and pumped back up again. It may be used directly in a district heating system. Ground-source heat pumps may also have boreholes, but they do not use geothermal energy. These boreholes may be up to 100m deep.

Figure 7.19
Schematic diagram
of a horizontal loop
ground-source heat
pump. An air-source
system would be
similar, but without the
underground loop;
instead, the left-hand
heat exchanger would
take heat from the
outside air drawn into
the unit.

Source: Author, with
thanks to Energy Saving
Trust

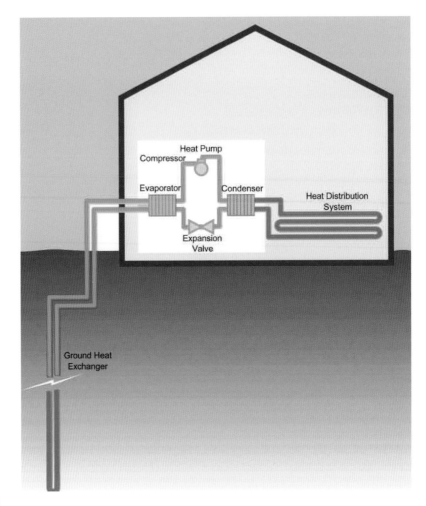

Figure 7.20 A trench
with a collection coil
containing the water
for a ground-source
heat pump.

Source: John Cantor

Heat pump manufacturers' estimates of their COPs should be treated with
caution, because real operating conditions will not reflect test conditions. The
standard used to test and quote for most packaged heat pumps is BS EN 14511.
This, for example, specifies test conditions of 7°C (44°F) outdoor (source) air
temperature for air-source heat pumps and a return and flow temperature of 40°C
(104°F) and 45°C (113°F) respectively. It is quite possible for air-source heat
pumps to use more energy than the heat source they are replacing, if specified
and installed incorrectly.

Heat pumps can transfer their heat to air or water. If water, this is to the
underfloor heating system. If air, a condenser indoors heats the air at the point
where it is supplied to the building. The filtered, pre-warmed air is directed into
the building from vents by a ground floor wall. One advantage of air-destination
heat pumps over the water-destination variety is that air into which the heat is
passed typically has a lower temperature (called the sink temperature) than that
of water. This results in a higher CoP and increased heat output.

Solar water heating

Solar water heating (solar thermal) works well in areas with east-southwest-facing roof space (in the northern hemisphere). It is often combined with other sources of heating, such as biomass and heat pumps. The storage tanks permit multiple inputs from the different sources of heat, which then transfer their heat from closed loops in their respective systems.

On average, throughout the year, around 60 per cent of water-heating requirements can be met this way in medium latitudes and 40 per cent in higher latitudes (depending on collector area and system efficiency). Closer to the Equator, almost all water can realistically be solar heated throughout the year. As long as the temperature in the collector is higher than that of the incoming cold water (usually about 10°C (50°F), then a solar water-heating system will save energy.

Some companies supply dual systems, which provide space heating as well as hot water. The solar collectors have to cover a much larger area. They supplement the use of another fuel source to heat spaces during the autumn and spring. Depending on how well the building is insulated and its location, a solar thermal system can supply 20 to 30 per cent of the heating demand. If combined with passive solar building design, it is possible to supply all of the heating requirements from the sun.

Sealed, indirect solar systems form the majority of industrial installations. They typically use the more efficient evacuated tube collectors, which consist of long glass tubes containing a metal strip inside a vacuum that absorbs the heat and transfers it to a liquid at the manifold end. These units are modular, with the advantage that the system can be expanded and installed quickly. A new type of solar collector, which is easy to install and up to 50 per cent cheaper, is polymer based. They are made from a single piece of moulded plastic.

Another type of solar collector, frequently but not always ground based, consists of rows of parabolic reflectors which, tracking the sun throughout the

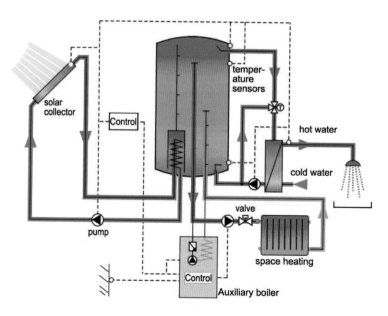

Figure 7.21 A simple solar water-heating system. The tank permits inputs both from the solar collectors and an auxiliary boiler, which could be gas, biomass, or a loop from a heat pump. The outputs are for space heating and hot water.

Source: Author

Figure 7.22
Evacuated tube collectors on an apartment block in Wezembeek, Belgium. They produce around 6.6 megawatt-hours (MWh) of heat energy per year.

Source: IEA-SHC (International Energy Agency-Solar Heating and Cooling programme)

day, concentrate its light on to a tube positioned at the focal point containing a similar liquid. Again, this is transported to manifolds and then to the heating system. This type of collector is only suitable for hotter climates where there is significant direct solar radiation.

Whatever the system comprises, it will need to be sized correctly by a professional contractor. Most professionally installed systems come with a ten-year warranty and require little maintenance, apart from occasionally cleaning the collectors, a yearly check, and a more detailed check every three to five years. In temperate and subarctic zones, systems are designed so that in summer they can completely meet all the water-heating requirements. In winter, most water heating will need to be supplied by another source, supplemented by solar thermal whenever it is sunny.

Solar space heating

Solar space heating uses some kind of solar collector and, optionally, some form of heat storage, plus a heat distribution system. The challenge with solar power for space heating is that when it is required it is usually winter and there is not so much sunshine. Nevertheless, with well-insulated buildings, especially those with an HVAC system with heat recovery (MVHR), solar space heating is possible, usually as a supplement to other forms of heating.

Air-based systems

The simplest system heats air drawn directly into rooms by natural ventilation or forced ventilation, using fans. They are really only appropriate for climates

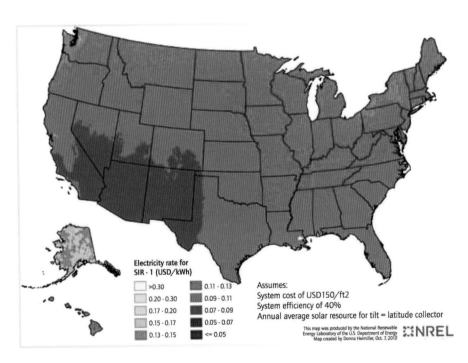

Figure 7.23 The viability of solar hot water systems compared to electric water heaters in the United States.

Source: NREL 2011

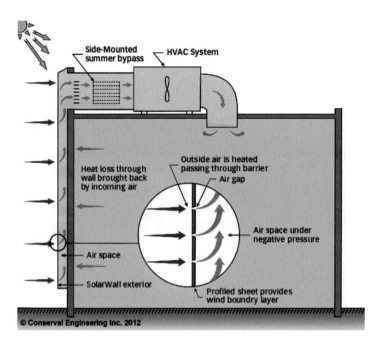

Figure 7.24 Cross section of a transpired solar wall integrated with a HVAC system.

Source: Conserval Engineering

with long, cold winters that have many sunny days. The solar heating air, heated using a glazed system with a large collector area, is allowed to enter the occupied space slowly and continuously from several different points. The most efficient operating temperature is around 32°C (90°F).

Sometimes, collectors are mounted on a roof, in which case they are connected into a mechanical ventilation system with heat recovery (MVHR/HVAC). This preheats the fresh air that is input into the ventilation system. If the incoming air is warmer than the recovered heat from the outgoing air, the recovered heat is not needed and the heat exchanger is bypassed. At night, and on colder days, the incoming air from the solar panel has the recovered heat transferred into it. Because the air's moisture content is normally low when the sun shines, these systems help to dehumidify the internal climate.

Collectors may also come in the form of cladding on the sun-facing exteriors. One design has glazing over a black metal plate that heats air, drawn from within the building at floor level, as it is drawn up by convection or a fan between the two layers, before it is directed back inside the building at ceiling level.

A further design, the transpired solar wall, consists of metal sheets perforated with thousands of tiny capillaries, each absorbing the sun's heat. It is positioned eight inches (20cm) away from the inner wall. Incoming air is then drawn through the holes, up the space in between, where it warms, and through ducts into the building, where it is discharged around the floor level of the lower floor. It rises through the building using the stack effect. Alternatively, it can be drawn into a rooftop MVHR/HVAC system. Ducting transmits the heated air around the building.

In a third design, soon to be commercially available, steel cladding is coated with dye-sensitised (DSC) solar-absorbing paint which is able to generate

Figure 7.25 Cross section of the transpired solar wall principle.

Source: Tata Steel

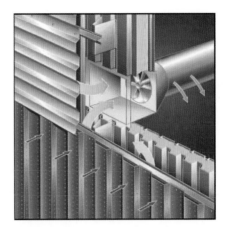

electricity. All these cladding formats are suitable for large industrial and commercial buildings both new and refurbished with cladding. Further exciting coatings are being developed for this type of cladding, including for heat production, carbon capture, anti-pollution, light enhancement, cooling and energy storage (at the SPECIFIC project at Swansea University, Wales, among other establishments).

Water-based systems

Water-based solar space heating systems are commonly hybrid or combination systems, preheating the water so that another heat source has less work to do, as described on pp. 145–6. Even in northern latitudes like Sweden or Canada, it is

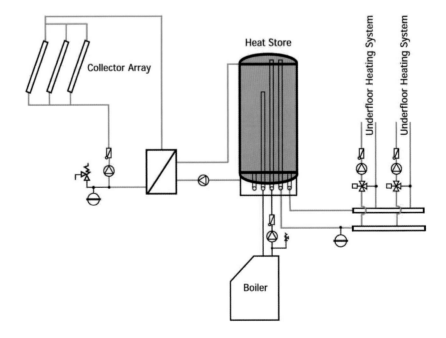

Figure 7.26
A schematic diagram for a heating system using a combination of a boiler and a solar collector array to supply underfloor heating.

Source: IEA-SHC

possible to halve primary energy use with such a system. The challenge is to size it correctly, so that the panel collector area and the storage tank volume match up at the most efficient level. This depends on local weather and latitude, the available roof area, and the pattern of use of hot water and building occupancy. The larger the solar collector area, the faster the storage tank will heat up; but too large and collected heat will be wasted.

Space heating from solar water and from heat pumps is at its most efficient when delivering heat at a low temperature to underfloor heating systems, as may be seen from Table 7.2.

Table 7.2 The delivery temperature of the heat required to effect the same degree of subjective comfort, using different distribution systems. Note that the higher the temperature to be reached the more energy is required disproportionally, and the larger the surface area of the heat emitter, the lower its temperature will be

Distribution system	Delivery temperature (°C)
Underfloor heating	30–45
Low-temperature radiators	45–55
Conventional radiators	60–90
Air	30–50

The concrete in the floor acts as a thermal store and maintains an even temperature, provided that it is well insulated. Figure 7.27 shows a typical schematic layout. Sometimes it is possible to dispense with a thermal storage tank if it is deemed that the thermal mass of the concrete floor is sufficient, once heated, to maintain a reasonable and consistent temperature throughout the year, day and night. An insulation layer with a minimum depth of 23cm (more than 9 inches) should be installed underneath the concrete floor, and thermal bridging minimised throughout. One example is a warehouse with integrated office for Neudorfer,

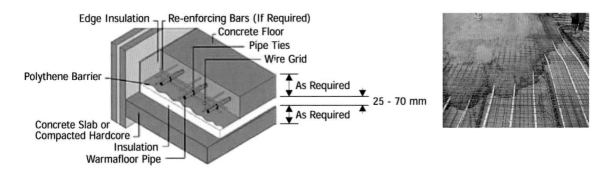

Figure 7.27 Installation of an industrial-scale system, as used in hangars and factories, based on the schematic diagram above. An array of solar collectors on the roof feeds an optional, suitably sized storage tank supplemented by a boiler. This serves an underfloor heating system, where the pipes are laid in tandem with a reinforced concrete wire grid. The instance in the photograph is of an aircraft hangar.

Source: IEA-SHC

in Rutzemos, Austria, built in 2005 (Figure 7.29). The section of the building containing the offices is built according to Passivhaus standards with an annual space heating consumption of 18kWh/m². Solar thermal collectors installed on the roof feed into an underfloor heating system, while façade-integrated PV panels supply electricity.

Heat stores

Whatever the heating source, there may be benefit from installing a heat store. Heat stores (also called heat accumulators) are large, insulated tanks that can store large quantities of heat in the form of hot water. They can be very useful additions to heating and cooling schemes, decoupling the direct link between the production of electricity and heat and its supply. Although often used in combination with CHP, heat stores may be coupled with any form of heat generation technology, such as a heat pumps or biomass boilers.

Solar cooling

It is possible to use solar power for cooling applications such as air conditioning. This is a good match of technology to purpose, since the more sunshine there is, the more requirement for cooling there is. The output of solar thermal chillers ranges from 10kW to 5MW, and can provide cold water as well as air conditioning. There are several techniques in development for solar cooling: absorption, adsorption, and solid and liquid desiccant cooling. What they all have in common is that the external energy required to drive the process can come from solar heat. With absorption chillers, which cover the 15kW to 5MW range, a heat-driven concentration difference moves the refrigerant medium (usually water) from the evaporator to the condenser. The market availability of absorption chillers is mainly applied in combination with district heating or heating from cogeneration. Their coefficient or performance (COP) varies between 0.7 and 1.1.

The 50 to 400kW range is generally served by adsorption chillers operating with a solid adsorbent. The input temperature of about 60 to 90°C (140 to 194°F) is provided by flat-plate or vacuum tube collectors with a coefficient of performance of 0.5 to 0.7. They cost considerably more than absorption chillers, and need about 3 to 3.5m² of collector surface per kilowatt of cooling capacity.

Desiccant and evaporative cooling systems achieve cooling capacity in the 20 to 350kW range. The operating temperature is only around 45 to 95°C (113 to 203°F); this means that the heat may be provided by simple flat-plate collectors and in some cases even air collectors, with a COP of 0.5 to 1.0 (the higher the COP, the better). The latest developments are triple effect absorption chillers with COPs of over 1.8. These systems supply cooling and dehumidification, and are widely available, with a high up-front cost but low operational costs.

In 2011, worldwide, about 750 solar cooling systems were installed, including installations with a low capacity (<20kW). More recently, a number of very large installations have been completed or are under construction, such as the system at the headquarters of the CGD bank in Lisbon, Portugal, which has a cooling capacity of 400kW and a collector field of 1560m²; and a system installed at the

Figure 7.28 A winery in Tunisia, which uses solar refrigeration. The solar collector at the front drives the absorption refrigeration machine. The wine in the fermenter tanks in the background is cooled by a cold accumulator.

Source: © ISE

United World College in Singapore, completed in 2011, with a cooling capacity of 1470kW and a collector field of 3900m². This installation is reportedly fully cost-competitive. It was implemented with the help of an energy services company (ESCo) model, under which the customer is not exposed to equipment or project costs; instead, the ESCo sells the resultant cooling capacity to the customer. The flat-plate collectors are run at a temperature of around 100°C to guarantee the cooling unit has a nominal working temperature of 88°C. A transparent Teflon sheet between absorber and glazing reduces heat losses at high temperatures.

Some systems use a cold store, where cold water is stored to be used when solar power is not available to power the chiller. Solar cooling works well with a heat pump, as the heat may be returned to the ground or air. The heat pumps may then be used in reverse in the winter, to heat the property.

HVAC systems

Having considered the components of heating, ventilation and air conditioning or cooling solutions separately, we are now in a position to consider the relative merits of integrated systems. Heating, ventilation and air-conditioning (HVAC) systems are integrated mechanical systems for maintaining an even temperature throughout the year in a building. HVAC systems may be purchased as a complete turnkey solution, although this may not be the best option, but they typically contain the following basic components:

Figure 7.29a and b Solar collectors for space heating on a Passivhaus standard office building in Austria.

- A source of heat, which supplies either air or hot water;
- Cooling equipment which produces cold water over which air is blown to be ducted to where required. This will eject heat;
- Pumps;
- Fans, to extract stale air;
- Heat exchanger for passing the heat in the outgoing air to the incoming fresh air;
- Controls, sensors and feedback loops if the process is automatic; additional manual controls.

Figure 7.30
Components of a typical small heating, ventilation and air-conditioning system. The unit incorporates a heating source, which can be of any type.

Source: Author

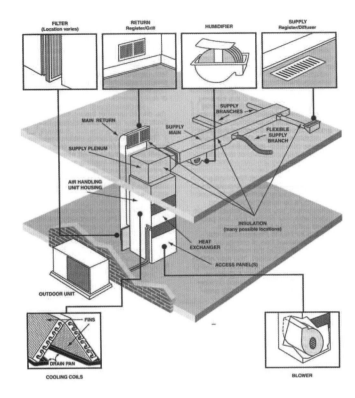

Having determined the degree day values throughout the year using the method above, this information is used to size the system. In fact, with some solutions it may be input into the controls for the HVAC system, or factored into its usage in another way. Many types of HVAC system are in operation, and an engineer is usually responsible for them either on-site or under a service-level agreement.

Most faults with HVACs occur due to poor design, inadequate specification, poor installation and inadequate maintenance. The energy manager should look for the following issues where opportunities may arise to optimise performance for efficiency:

- Thermostats should not be located near doors and windows. They must be set at the right temperature, usually 18 to 20°C. Typically they can be set too high. It may be necessary to talk to staff to rectify any misconceptions that a higher set temperature means that, for instance, rooms warm up quicker (it doesn't).
- In offices and similar environments where heating is controlled manually, thermostatic radiator valves (TRVs) should be fitted to individual radiators, which provide the greatest degree of manual control. However, they should not be located near thermostats, as their operation can interfere with each other. They should also be tamper-proof and locked at a fixed setting, to prevent staff from interfering with them.

- The control systems can become desynchronised from the building's operational requirements. Timing schedules may be set inappropriately for actual occupancy patterns, and even set for the wrong time, date and season. In some cases start-up and switch-off times are programmed into the system, which can make it more difficult to modify. Identifying and rectifying the settings will immediately produce energy savings.
- Pumps and fans may be incorrectly specified. The energy consumption from HVAC applications such as fans and pumps rises with the cube of the flow. Often they run at a fixed speed, which is frequently not matched to the requirement. Fitting variable-speed drives to these motors allows their speed to track the required demand, and is highly recommended to achieve considerable financial and energy savings: energy savings of between 5 and 48 per cent are possible. Sophisticated drives come with a range of energy consumption monitoring and control systems. Drives may even be purchased as a series of modules with optional plug-ins such as a fieldbus for integration with the BEMS (building energy management system), a local control panel, mains disconnect and so on.
- Frost-protection set point temperatures on heating systems may be set too high, which will cause them to run needlessly when the building is not occupied.
- In air-conditioned buildings it is often found that heating and cooling systems are set to run simultaneously, in conflict with each other, and perhaps independently of the weather. They should be set so that there is at least 5°C (9°F) between when one cuts out and the other kicks in.
- Often, two or more boilers are servicing the same space, together with chillers or cooling towers. It is worth checking, at any given time, especially in times of non-peak demand, that the minimum number of boilers and firing times are being used for the purpose. When not in use they must be isolated from

Figure 7.31 Variable speed drive-controlled HVAC systems come in a variety of sizes.

Source: ABB

the system, as otherwise they can still absorb heat and waste energy. Proper sequence control avoids short cycling, where the boiler keeps firing to top up the system. Boilers should be inhibited from firing up when not required. Rotation of the boiler firing order evens out wear, prolonging the life of the system. If a condensing boiler or a combined heat and power (CHP) unit is present, it should always take priority. These should be sized to provide the base load.

- Systems should be regularly inspected for defective dampers, valves and actuators as part of the maintenance schedule, and they should be replaced at the earliest opportunity.
- Any control system relies on the sensors sending accurate and pertinent information. They should therefore be situated in the most relevant place, and be calibrated correctly. Sometimes this involves conducting independent measurements to verify accuracy. A portable data logger would be necessary. The need for this may be identified by noticing that the heating system is systematically keeping a room at the wrong temperature.
- Sometimes, where the HVAC system is incorporated within the BEMS, there is a complex series of different control loops, which affect both heating for water and space. This can result in uneconomical operation of the boilers. An audit of the BEMS will usually show up this and any other existing anomalies.
- All heating pipes and ducts should be properly insulated with no breaks.

If different parts of the building require heating or cooling at different times, or require different amounts of passive heating or cooling, then the building should be divided into different zones, each with its own system of controls and heating or cooling pattern, such as lower temperatures in unoccupied areas or different heating times. This applies especially to multistorey buildings, multiple buildings served by the same boiler house, shared buildings and multipurpose buildings. Before doing this, it should be checked that the installed equipment can cope with zoning. It is appropriate for large or complex buildings, but staff must be available to manage it.

Motorised valves are used to control water flow from the boiler to the heating and hot water circuits. Two-port valves may be used to provide zone control. Three-port valves are used with controls such as weather compensators. Thermostatic

Figure 7.32 Well-insulated piping in an HVAC system.

Source: Public domain (Steve Karg)

radiator valves (TRVs) can also assist with zoning. For larger buildings and sites, separate pumps and pipework would be required. A chief benefit is increased staff comfort and productivity.

Much of the above applies to wet heating systems. Warm air systems are relatively simpler. Controls mainly include a programmer, room thermostats, temperature sensors and controls on the heater, controlling its firing and fan speed. Variable speed motors vary the fan speed to match the firing cycle of the burner, producing large energy-saving gains. It should be a priority to replace any fixed-speed drives. Air systems should be linked to the air-conditioning system with heat recovery from the exhaust air.

Variable refrigerant flow

A variable refrigerant flow (VRF) system is designed to minimise efficiency losses found in conventional HVAC systems. It may be thought of as a hybrid system incorporating a heat pump. It is based on the principle that a heat pump system may be designed to cool the building too if allowed to run in reverse. VRF uses a refrigerant as the cooling/heating medium instead of water, and allows one outdoor condensing unit (the coil in the ground or in the air) to be connected to multiple indoor fan-coil units, affecting different spaces, which are individually controllable by the user. By operating at varying speeds, they work only at the rate needed, thereby saving energy. They may additionally provide heat recovery, whereby heat rejected by units in one area that needs to be kept cool can provide heat to an area that needs to be heated.

Thermal destratification[2]

Thermal stratification is caused by hot air rising up into the ceiling or roof space because it is lighter than the surrounding cooler air. It happens in all buildings and can create dramatic temperature differences between the ceiling and floor of a room. Where they are fitted, this means extra work for HVAC systems to maintain building interiors at a set and even temperature. A destratification fan will capture and reuse this heat. Mechanical engineers will insist that good HVAC design will ensure no stratification occurs within a space; however, anyone who has worked in the roof space of a building will have found this to be inaccurate.

Destratification fans offer significant energy and monetary savings, optimise HVAC system output, improve environmental control and comfort levels, and act as a cost-effective solution to reduce a building's carbon footprint. Destratification is also effective in non-air-conditioned spaces such as warehouses and factories, where internal environments can be cold during winter months, even though heat generated by machinery, processes and people can be present in the roof space. This heat can be recirculated to cost-effectively improve temperatures at floor level. The key to efficient destratification is delivering continuous, direct, non-turbulent airflow with maximised throw distance so that high- and low-level temperatures

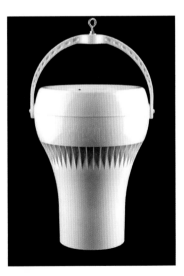

Figure 7.33 Modern thermal destratification fans direct hot air from ceilings to where it is needed with minimal electrical usage. New models may be wirelessly controlled, making installation easier.

Source: Airius

are fully mixed in a controlled fashion, using low-energy motors. Fractional amp draw axial turbine fans are recommended for this purpose due to their low power draw and efficiency. The leading brands begin operating on only 12 watts and may be angled up to 90° off vertical. This means that the flow of air may be directed to where it is required. Standard fans generally use more energy and cannot push air in a single direction for any distance beyond 3 metres, instead mixing the air.

In the UK, major retailers such as Morrison's, Tesco, Sainsbury's and John Lewis are using thermal destratification systems, saving significant amounts in operational heating and cooling energy. They can improve freezer and chiller aisle comfort levels with no impact on fridge or open freezer cabinet performance. Morrison's now specify thermal destratification in all stores. The United States Navy has conducted research[2] which found a 40 per cent reduction in energy consumption in two facilities implementing thermal destratification fans. Short payback periods, of well under one year, have been reported.

Case study: Ebök GmbH office building refurbishment, Germany

Figure 7.34a and b The Ebök GmbH office building, Germany, a former barracks, before and after refurbishment.

Source: Forschung für Energieoptimiertes Bauen

This was the first building worldwide to be awarded a Passivhaus certificate for its refurbishment, in Tübingen, Baden-Württemberg, Germany. A problem frequently encountered with the renovation of existing buildings is how to insulate the ground floor. Deeper thermal insulation thickness laid on the concrete foundation slab would have necessitated raising the door lintels. Instead, increased insulation was added to the building's outer walls. With the help of dynamic, two-dimensional heat flow calculations, it was shown that this, combined with reduced floor insulation, enabled the temperature trapped in the ground beneath the building to increase over time, creating a heat sink, which reduces heat losses and leads to higher temperatures at the base. Apart from this, thermal bridges were eliminated.

Air conditioning and ventilation are achieved solely via mechanical air supply and exhaust systems, and, if required, by opening windows. A brine-air heat pump preheats the supply air and provides frost protection. The brine comes from a flat-plate collector laid in the ground around the building. Run in reverse, it cools the air as required. Free night cooling is implemented with support from the ventilation system (mechanical ventilation).

For structural reasons, the roof extension had to be built using lightweight materials. The ceilings of the top floor were therefore clad with a phase change material: plasterboard impregnated with micro-encapsulated paraffin.

At 20kWh per square metre per year for heat (thermal heating and hot water), and approximately 7kWh per year for lighting and building services equipment, the building requires only 15 per cent of the primary energy used by typical existing office buildings. The building has been monitored for some years and found to perform satisfactorily. Its payback for the work was 26 years, but, since it was first of its kind, this has now been significantly reduced for modern refurbishments.

Case study: Sheikh Zayed Grand Mosque in Abu Dhabi

This mosque is the third largest in the world and iconic because of its triple-domed roof and multi-domed structure. It is a large, open and busy space in a very hot part of the world.

Thermal simulation and analysis of the building was conducted to determine what level of thermal comfort would be required, and to optimise HVAC system

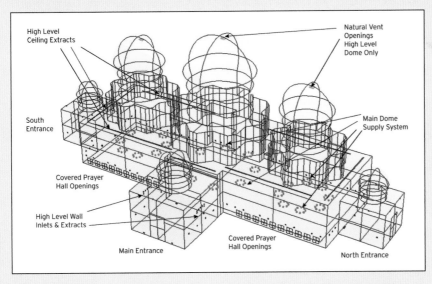

Figure 7.35a and b A design sketch and photograph of the Sheikh Zayed Grand Mosque in Abu Dhabi.

Source: (a) IES; (b) Photo by Max Montagut

Figure 7.35a and b continued.

performance. A building information model (BIM) captured the structure and thermal aspects of the building. Software allowed the geometry to be created from scratch as in a conventional architectural CAD. The model was populated with thermal information and HVAC data. This included data on the occupancy levels, fabric structure and internal equipment behaviour. An initial model determined the impact of a conventional air-conditioning system. This was then refined with various options, including a fresh air supply from the central air-conditioning induction system. The combined effects on HVAC loads, energy consumption and thermal comfort were analysed for each option, allowing the optimum design to be determined. A computational fluid dynamics (CFD) analysis of the main prayer hall, which can accommodate 9000 people during Ramadan, was carried out to model thermal comfort and airflow patterns in more detail. The distribution from the high-level air-conditioning induction system could be visualised in the model. In addition, the effects of extract and supply louvres located at high level around the central support columns could be determined.

Case study: a district heat metering and billing system, UK

A metering and billing system was created for 1036 homes spread across 28 housing schemes that are connected to the community heating supply across a local administration area in West Yorkshire, England. Kirklees Neighbourhood Housing (KNH, a not-for-profit company that manages Kirklees Council's council housing) commissioned the £1.7 million contract, which includes a five-year

Figure 7.36 A district heat meter in situ.

Source: KNH

maintenance and data administration programme, from energy metering and billing specialist ENER-G Switch2 (ES2). This cost will be recovered over a long contract period as part of a fair service charge to residents.

In the first six months, residents have reduced overall energy consumption by 15 per cent; when winter consumption is factored in and residents get used to the system, KNH predicts that the annual average consumption will reduce by 30 per cent to 50 per cent. The system uses data collection and energy management technology for both pay-as-you-go and credit billing consumers. It incorporates an in-home display that lets customers see, in graphical form, how much energy they are consuming, when, the cost, and how much credit they have. Additional meters may be connected for potable water and electricity.

The pay-as-you-go element works much like topping up a mobile phone, using smart secure wireless GPRS technology to replace the traditional token-based prepayment system. There is a text alert facility to advise customers of low credit or to communicate service and energy-saving messages. The system avoids any requirement for an expensive building management system to be installed, or Mbus or other hardwired system.

Analysts continually monitor consumer data and can identify exceptional usage patterns or instances where residents frequently opt to disconnect their supply. This serves as an early warning signal that can assist operators in targeting support towards residents who may not be using the system effectively. In one case, ES2 identified that an elderly resident's heating consumption had doubled over a 48-hour period. Staff established that he was immobile and were able to arrange immediate support.

'The key to reducing energy consumption is being able to see what you are using,' said Barry Goodwin, KNH Project Manager. 'With the economy of scale of supplying heat via a district heat network, we already provide lower-cost heat, but now that residents can track how much energy they are using and are taking action to reduce consumption, annual heating bills, including winter consumption, could average £5 per week. Heating bills for a similar sized property would cost approximately 60 per cent more if residents were purchasing gas or electricity from one of the big suppliers.'

ES2 designed and developed the system, with specialist software support provided by ADI, and hardware design provided by Blueprint.

Notes

1 *Final Report: In-situ monitoring of efficiencies of condensing boilers and use of secondary heating*, Energy Saving Trust, June 2009.
2 Thermal Destratification Technology at West Bethesda, MD, April 2010 (International Energy Agency).

Energy managers share their experience

Samantha Dean, Strategic Partner Channel Manager at IMServ

Sam heads up the partnering sales and strategic collaboration channel within Invensys IMServ. Sam's experience includes 12 years of expertise within the energy sector, working for FTSE 100 organisations such as National Grid, Rolls Royce, Siemens Metering and Invensys IMServ within Leadership, Business Development and Head of Sales roles. She is educated to degree and MBA standard with an understanding into the current market trends and demands to deliver a customer-centric approach to Monitoring, Control and Data Intelligence.

What do you recommend energy, building or facility managers to tackle first?

The number one priority is to ensure selection of the most appropriate energy supply package. Not all business requirements are the same, with different views on long-term stability and risk, affecting energy supply and management choices. Businesses can typically save up to 17 per cent overall by implementing a hands-on monitoring, visualisation and control energy focus.

My experience has taught me the seven other most important energy-saving tips for energy management:

1 Appropriate lighting: Simple changes such as the addition of LED lighting to office buildings can deliver significant energy and cost savings. Businesses could also install light sensors to control usage, this works in a similar way to street lighting by switching on and off automatically, fading and getting brighter. Most importantly switch off all unnecessary lights when staff leave the office, a simple cost saving tip that could save approximately 15 per cent on average.

2 Meter energy-consuming 'hot spots': Businesses can't monitor what they can't measure, making transparency of metering data both powerful and effective. Sub-metering energy monitoring systems enables businesses to accurately see what their energy usage is and where. This can enable correct apportionment of energy to tenants or cost centre owners if required, reducing consumption effectively through awareness.

3   Utilise Building Energy Management Systems (known in the market as BEMS or BMS): These systems monitor and control the building's mechanical and electrical equipment to ensure energy is being used efficiently and effectively. Changes to in-house controls and environments are made to ensure that all systems are working at an optimum level. In addition, systems supported by a proactive Energy Bureau can maintain employee engagement, bringing substantial benefits and savings. Support of a Bureau Team will also ensure that energy surges, spikes and patterns in usage are monitored on the customer's behalf.

4   Energy performance contracting (EPC): An internationally recognised, innovative financing mechanism that uses cost savings from reduced energy consumption to repay the cost of installing energy management measures. This offers considerable benefits, allowing energy savings to be achieved without upfront financing.

5   Empower employees: The simplest way to change attitudes is presenting your employees with hard facts. By utilising innovative and dynamic dashboards, an interactive visual tool that enables the display of your energy and data in a user friendly and easy-to-understand format, you can see the over-usage points and also shout about your energy- and cost-saving successes. This is the key to educating stakeholders and employees, raising energy awareness and changing behaviour.

6   Heating: Substantial savings can be made by ensuring your heating systems are working efficiently, and maintained correctly. There are a few simple tips that are worth remembering. Don't overheat the building; by lowering the temperature by just 1°C your annual heating bill could be reduced by up to 8 per cent. Have your heating system regularly maintained and replace inefficient boilers.

7   Cooling and ventilation: Alongside heating, cooling and ventilation can account for a major proportion of spend on energy, so simple steps such as considering natural ventilation before switching on the air con can make all the difference. Look to change behaviour through education, make the energy- and cost-saving requirements known and people will be more likely to adapt their behaviour.

8

Minimising water use

Most businesses and organisations use a great deal of water. This has an impact upon energy consumption and carbon emissions through activities such as pumping, heating, purification and treatment. Water is also becoming increasingly scarce and expensive. For all these reasons, facility and energy management duties often include water management.

In the UK, water contains an average of 1.47kg of embodied carbon dioxide-equivalent per 100 litres, through sourcing, transportation and heating. The embodied carbon contained in water supplies elsewhere will depend on the proportion of fossil fuels used in grid-supplied electricity in the local area. Capturing and recycling used water can reduce water use by millions of gallons/year, as well as save money.

Conducting an audit

Just as with energy, an audit of water use is the place to start. This will help identify very simple ways to easily capture and reuse water without extensive capital investment. The audit should cover all the sources of water use. It will consist of a list in spreadsheet form of water-using outlets, their function, flow rate and use per week or per day. Comparing this to utility bills or readings from a building management system can reveal discrepancies and targets for action. The audit should also quantify the true cost of water to the organisation.

Flow rates can be measured by timing how long it takes to fill a container with a known capacity. Cistern volumes may be calculated by tying up the ball-cock and refilling the cistern from the same container. These volumes may then be multiplied by a reasonable estimate of the number of times they are used per working day throughout the year. The end result should be a table with headings similar to the following:

Item	Location	Number of units (A)	Flow rate (gallons or litres/minute) (B)	Operating time (minutes/day) or (uses/day) for each unit (C)	Water used (gallons or litres/day) = A × B × C	Comments

The penultimate column may then be added up to determine the weekly/daily/monthly total usage. While doing this, it is a good opportunity to discuss with staff how they use water and ways to minimise use.

Water use can be checked against bills. Bills identify water use by volume. The average office water use is 4 cubic metres per employee a year.[1] This enables water use per employee, in an office, to be benchmarked by dividing the annual water use (in volume) by the number of staff (full- and part-time). For example, for an office with 36 full- and part-time staff and a six-monthly water use of 101m^3 (taken from the bill) this would be:

$$101 \times 2/36 = 5.61 \text{ m}^3/\text{employee/year}.$$

This signifies that there is room for at least 20 per cent savings.

A water-efficiency strategy

The next stage is a water-efficiency policy and strategy. Most organisations already have a water management strategy; if so, it should be reviewed annually. As with electricity and energy use, the single most important aspect of water use is to minimise it. If the demand is minimised, the energy and financial cost of supplying it is reduced.

A good water use policy should:

- set a target for reducing water use in buildings, with further, more stringent requirements for all new-build and major refurbishment projects;
- embed the targets within corporate policy and processes;
- set corresponding requirements in project procurement and the supply chain;
- measure performance at a building level relative to a corporate baseline and report annually on overall corporate performance.

The units or key performance indicators (KPIs)[2] chosen may be any of the following:

- volume of potable water consumed: gallons or m^3 per year, with reference to the number of full-time-equivalent occupants (commercial spaces), number of visitors (retail, leisure or public spaces), number of residents (housing);[1]
- square feet or m^2 Net Lettable Area per year;
- % reduction in use relative to 20xx.

Minimising water use

The water minimisation hierarchy in a nutshell consists of the following priorities, in order:

1 Eliminate wasted water.
2 Improve efficiency and use alternative sources.
3 Reuse water.
4 Recycle water.

Initial steps to minimise use may be obtained by:

- having an organisational water-efficiency policy;
- installing metering and sub-metering, perhaps in conjunction with a building energy management system;
- installing water-efficient components;
- replacing potable water with water from other sources (e.g. rainwater or greywater) where appropriate;
- minimising the energy and carbon emissions associated with the generation, storage and supply of hot water (within the property);
- influencing user behaviour through water system design;
- regular checking and maintenance to ensure that water-consuming fittings, appliances, controls, pressure/flow regulation and monitoring systems are adequately maintained and work safely and in line with their design performance;
- having a targeted replacement or retrofitting programme for fittings or appliances.

Water meters alone very often have the effect of reducing use by up to 17 per cent. The energy manager should prepare a financial case for action based on the estimated capital and life cycle costs of any work required and the value of reduced water and energy use. Consideration should be given to the timing of any investment so that it matches existing plans for expenditure. As with an energy efficiency action plan, a water efficiency plan may address the following factors:

- The project/building's current or projected installation of water-consuming fittings and appliances and their current or likely use.
- The objective (e.g. maximum consumption level, or corporate target for improvement).
- Forecasts of alternative building outcomes for end-use water consumption arising from the use of components with different practice levels of water efficiency (estimates should be on a per person or per m² basis, and overall).
- A building-specific target for potable water use (design intent or in-use), and/or water-use specifications for different types of fitting and appliance, that meet or exceed the client requirement.
- Projected financial, water and energy savings and associated financial costs from going beyond the minimum.
- A log of designed-in and actual water use (recorded over time), supported by evidence of actions taken.
- A procedure for monitoring and review of performance against the target, together with a timetable for updating the water efficiency plan and capturing lessons learned.
- Allocation of who is responsible at each stage of the building life cycle (inception, design, construction, use) and who is responsible for implementing the actions.

Universal checks

The following apply to all types of establishment, including hotels and offices.

Checking for leaks

Checking for leaks should be done regularly. One way to do this is to take a meter reading last thing at night and again first thing in the morning, or when it is not expected that any water should be used. If the reading has changed it may be due to a leak. Daily or weekly logging of water meter readings and working out how much average water is used for each purpose can yield a benchmark from which to improve. Washers in dripping taps must be replaced immediately; they can waste at least 5500 litres of water a year. Leak monitoring equipment should be installed if possible and auto shut-off of flow to toilet areas when unoccupied.

Valves are a frequent source of problems: faulty valves causing tanks to overflow continuously can cost a fortune in water and effluent charges.

Comparing the results of meter readings to water bills can reveal discrepancies, such as more water being billed for than is actually metered. This means that a thorough check should be undertaken. Sometimes it is the case that nothing obvious is found. A leak detection subcontractor should then be called in, since sometimes leaks occur in underground pipes.

Figure 8.1 Leaks can waste a fortune in water.

Source: WRAP

Table 8.1 High-efficiency target levels for water use

Fitting/appliance	Ideal water-use target (at pressures up to 5 bar)
Shower (mixer) ≤	6 l/min
Shower (electric)	6 l/min
WC ≤	3.5 l/flush (single flush or effective flush)
Urinal ≤	3 l/bowl/hour during building occupancy with user-presence activated flush, 0 l/hour outside of occupancy and activation period, with minimal water use in maintenance. Or fit waterless urinals
Tap (basin) ≤	4 l/min
Tap (kitchen) ≤	6 l/min
Bath ≤	155 l capacity excluding body mass within the bath
Washing machine ≤	7 l/kg dry load
Dishwasher ≤	0.7 l/place setting

Source: WRAP

Taps

A running tap can waste over 6 litres per minute. Spray taps can save about 80 per cent of water used by conventional taps. Some users complain that sinks do not fill quickly enough, in which case a 'Tapmagic' insert can be fitted to most taps with a round outlet hole or standard metric thread. At low flows, Tapmagic delivers a spray suitable for washing hands and rinsing toothbrushes. As the flow is increased it opens to allow full flow.

Tap controls are available in both new and retrofit versions, including infrared (PIR), battery-operated and simple push-top and spray taps. PIR sensors can be fitted to taps which detect when a hand is beneath and turn on automatically. This means taps are turned off when not in use. Basin taps should be limited to 4 to 6 litres per minute and sink taps should be limited to 6 to 8 per minute. All mixers should have a clear indication of hot and cold with hot tap or lever position to the left.

Toilets

Installing urinal flush controls typically saves around 70 per cent of water used for flushing. One organisation saved £1300 ($1800)/year in water and sewerage costs by installing a passive infrared (PIR) sensor. With an installation cost of around £960 ($1300), a payback of just over eight months was achieved.

System displacement devices can reduce the volume of water required to fill a system. Low- or dual-flush toilets may be considered if replacing existing units. Restrictor valves can be easily fitted to supply pipes to keep the water flow concept, regardless of fluctuations in water pressure. Waterless urinals are available and are becoming increasingly common. These use microbial treatment to self-clean.

Case study: Haulbowline Naval Base, Ireland

This base is occupied 24/7 all year round. An uncontrolled urinal cistern was flushing an average of four times per hour using 9 litres (16 pints) per flush. There were four of these cisterns on the base. This resulted in the following:

$$4 \times 24 \times 365 \times 9 = 315 \text{ cubic metres of water flushed per annum}$$

The Department of Defence paid for this water at a rate of €1.95 ($2.56) per m², giving an annual cost of €614 ($806). In order to minimise this expense, it replaced the urinal valves with more efficient ones which meant, with 18 flushes per day (high estimate), that just 59 cubic metres would be used per year. This cost €115 ($151) p.a., a saving of €499 ($655) per year. The retrofit cost was €150 ($200) per urinal (approx.), €42 ($55)+VAT for valves and the balance for labour, yielding a payback period of seven months.

The system was successful and so was then installed in all urinals in new Naval Service Combined Command Headquarters. Result: €3500 ($4600) savings per year and 1.79 million litres of water conserved (urinals alone). Push-button or PIR-controlled taps and low-flow shower heads were to be installed as standard in all new buildings.

Hot water systems

Premises with mains pressure hot water will tend to use more water than those with a gravity system due to the higher flow rates from hot taps and showers. However, with good design efficiency savings (and, for example, better showering) can be achieved with mains pressure hot water by fitting the following:

- small bore pipes (10mm (0.5″) is the smallest available);
- regulated aerators;
- low-water-use showers;
- pressure and flow regulators.

Figure 8.2 An aerated tap saves water but gives the feeling of a full flow of water.

Source: Wikimedia Commons (Nikthestoned)

Figure 8.3 Properly insulated boilers.

Source: Royalty Free Picture Agency

If the mains or header tank pressure is greater than 3.5 bar pressure, regulators should be installed to maintain a constant pressure, to limit pressures in mains-fed hot water systems and to prevent damage to fittings. These regulators are normally adjustable and maintain a constant flow independent of supply pressure, and may be fitted in-pipe or at each tap or shower. An alternative is a shower head or tap outlet with a built-in regulator.

All hot water pipes should be properly insulated and sited above cold water pipes to reduce heat transfer. A radial layout for pipes to outlets from the tank will also help keep heat losses down.

The bore (diameter) of a pipe has a great impact on energy wastage, as may be seen in Table 8.2. The narrowest available bore should be chosen and dead legs (the distance travelled between the boiler and the tap) minimised to avoid losing heat.

Figure 8.4 Using thermostatic control of individual radiators to make sure heat is only supplied where needed and to the correct level.

Source: Royalty Free Picture Agency

Table 8.2 Dead leg volumes

Pipe diameter	10mm plastic	15mm plastic	15mm copper	22mm plastic	22mm copper
Litres per 10m pipe run	0.6	1.1	1.5	2.4	3.1
Max length for 1.5 litre dead leg (m)	25.0	13.0	10.0	6.0	5.0
Length for 30-second wait with 1.7 litres per minute spray fitting (m)	14.0	8.0	6.0	3.5	3.0

Table 8.3 Typical flow rates, water use and $kgCO_2e$ (assuming the UK's 65 per cent fossil fuel gas/coal mix of grid) used by different types of showers

	4 litre/min (0.88gal/min)	7.2kW electric	9.8 kW electric	6 l/min (1.32 gal/min) water saver	9.5 l/min (2.09 gal/min) water saver	'Power shower'
Flow litre/min	4 l/min (0.88gal/min)	3.5 l/min 30°C temp rise (0.77gal/min)	4.7 l/min (1.03 gal.min) with 30°C temp rise	6 l/min regulated flow (1.32 gal/min)	9.5 l/min regulated flow (2.09 gal/min)	Typically 12+ l/min (2.64+ gal/min)
Water use for 5-min shower	20 litres (4.40 gallons)	17.5 litres (3.85 gallons)	23.5 litres (5.17 gallons)	30 litres (6.60 gallons)	47.5 litres (10.45 gallons)	60+ litres (13.20 gallons)
As a % of 70-litre bath	29%	25%	39%	43%	68%	>85%
$KgCO_2$ emitted during the shower	0.07–0.27kg (0.15–0.60lb) (gas boiler)	0.34kg (0.75 lb) (direct electric)	0.45kg (1lb) (direct electric)	0.27–0.4kg (0.60–0.88lb)	0.42–0.63kg (0.93–1.39lb)	0.53–0.8+kg (1.17–1.39lb) (electric)

Source: UK Environment Agency
Note: Units are UK gallons. To convert to US gallons multiply by 1.20095.

Most modern fitted showers unfortunately now use electric on-demand heaters. Ideally these should be avoided in preference to the most efficient combi boiler gas-fired heaters or tank-fed from renewably heated water, for example, solar water heating, a heat pump with a buffer tank, and/or a biomass stove. Adding an electric shower can potentially save water but increase running costs and CO_2 emissions. A mixer shower (with pump if the existing head from a gravity system isn't sufficient) with flow regulation provides a more carbon-effective solution.

Figure 8.5
Well-insulated hot
water pipes.

Source: Royalty Free
Thinkstock

The Water Technology List

The Water Technology List is a list of certified products in the UK that are among the most water efficient available. 'Enhanced capital allowances' may be claimed on their purchase. Visit: www.eca-water.gov.uk.

Rainwater harvesting

Rainwater may be collected, stored and used for toilets, washing machines, gardening and other purposes. It does not need disinfecting, merely filtering. The larger the roof, the more cost-effective the measure. In order to calculate the benefit, the annual rainfall for the location should be known, together with the roof area. This is then multiplied by a drainage factor which is dependent upon the roof type. The higher the roof factor, the greater the proportion of rain falling on the roof will reach the gutter and be collected. Since rainfall is sporadic, storage may be needed. The size of the tank should be sufficient to hold about 18 days' worth of demand, or 5 per cent of annual yield, whichever is the lowest.

Table 8.4 Approximate annual yield of rainwater in cubic metres per year for a range of roof sizes with varying rainfall

Plan roof area (m²)		50	75	100	125	150
Mm	500	15	22.5	30	37.5	45
rain	1000	30	45	60	75	90
per	1500	45	67.5	90	112.5	135
year	2000	60	90	120	150	180
Plan roof area (yd²)		**59.8**	**89.71**	**119.60**	**149.5**	**179.40**
Inches	20	17.94	26.91	35.88	44.85	53.82
rain	40	35.88	53.82	71.76	89.71	107.64
per	60	53.82	80.73	107.64	134.55	161.46
year	80	71.76	107.64	143.52	179.40	215.28

Table 8.5 Drainage factors for different roof types

Roof type	Drainage factor
Pitched roof tiles	0.75–0.9
Flat roof smooth tiles	0.5
Flat roof with gravel layer	0.4–0.5

Example: A roof has an area of 530m² (633.87yd²). It receives 1500mm (60 inches) of rain per year. The maximum collectable volume is therefore 90×5.3 = 477m³ (624yd³). It is a flat roof so this is multiplied by 0.5, yielding a total of 238m³ (311.29yd³).

To calculate the tank size, multiply this figure by the filter efficiency (90 per cent) and by the 5 per cent. This results in 10.73m³ (14yd³).

Figure 8.6 Three (5000-litre or 22,730-gallon) rainwater harvesting tanks supplying the False Bay Ecology Park centre in South Africa with all its fresh water needs. From the tanks the water is pumped into the building with the pumps on the right.

Source: Cape Water Solutions

A system will involve the following:

- reliable guttering (steel or copper);
- downpipes;
- accessible filtration;
- frost-protected storage away from sunlight at a temperature which prevents bacterial growth;
- a floating intake to draw water from the top of the water so sediment at the bottom is not collected (new water comes into the tank near the bottom);
- a clearly labelled separate system of pipes alongside the existing mains backup plumbing system to pump and direct the water to where it is going to be used within the house (the water should be oxygenated);
- a rat-proof overflow.

The pipework for this separate network of water supply should be clearly labelled throughout the system. Standards exist for this and should be followed.

Rainwater harvesting may be integrated with SUDS (sustainable urban drainage solutions). This means that, provided it is properly treated if pollutants are likely to be present, all surface water on a site can be potentially collected and reused. Surface water drainage systems would lead into a tank, from which the water may be pumped for reuse. The sizing of this tank would be calculated using a method that takes into account extreme weather conditions, such as once every 50-year occurrences, or have provision for overflow into balancing pond/swales. These would be open to visual inspection and periodic cleansing. Integration of the two systems, SUDS and rainwater collection, makes it a double win.

Greywater reuse

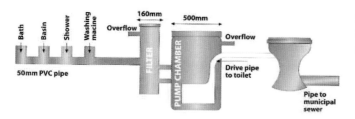

Figure 8.7 Basic components of greywater reuse system.

Source: Cape Water Solutions

These systems reclaim water from baths, wash-basins and showers, or other places where water is not likely to be too contaminated, store it and usually reuse it in flushing toilets. Savings vary from 5 to 36 per cent. Treatment is necessary because warm, nutrient-rich greywater incubates bacteria when stored. This means that greywater should also only be stored temporarily. Electronically controlled dump valves can control this by emptying tanks after a period of inactivity before refilling them with mains water. Chemical disinfectants such as chlorine or bromine compounds may be added. In larger systems, the greywater may be treated in a small sewage treatment plant, using membrane filtration or ultraviolet technology. This is normally not cost-effective unless such a system is already in place.

Greywater systems require more frequent attention. Filter cleaning may be required if self-cleaning filters do not always function. They do have the advantage of a more predictable and constant supply of water than rainwater systems.

With both rainwater and greywater reuse, the carbon cost of creating and installing the system, plus running it using the pumps, should be taken into account before a decision is taken. Any pumps used should be of the minimum power specification for the job and not oversized. If possible, gravity-fed systems should be deployed to minimise or remove the need for a pump. In both cases as well, using the water for irrigation purposes is the simplest end use.

Note

1 Some buildings will contain a mix of uses (e.g. office/staff areas and public spaces such as retail floors). In this instance, benchmarking should be based on the dominant use type (if clear) or independently for each type of space if each use is significant.

Energy managers share their experience

Robert Kelk, Energy Officer for Manchester City Council, UK

What does your job involve?

Energy monitoring and targeting (M&T), procurements, metering and smart metering, coordinating energy projects, invoicing audits and validation, data administration, and advanced scientific energy and financial modelling.

What qualifications did you gain?

Degrees and postgraduate degrees in economics, politics, energy management, Display Energy Certificates and energy procurement.

I had always been interested in the concept of climate change and issues related to the economy and sustainability. My degrees were also heavily involved with data and financial statistics and their analysis. So moving into energy management seemed like the perfect fit for me. It involves the analyses of energy and financial data but also has direct links with political and economic issues.

The job is continually expanding and of interest. Since I've moved into the area of energy, my role and knowledge has become much broader. A further degree (more directly related to energy management) has given me a more thorough mechanical and electric engineering understanding. I've also become very interested in electric metering – specifically smart metering – and how this can be used to reduce energy consumption and expenditure.

What does your day-to-day work look like?

I'm involved with energy procurement and energy M&T, but no two days are the same. I'm currently working on two main projects. First, electric smart metering. We're now on phase four of our electric smart metering roll-out programme. So far we have completed around 300 smart metering installations. We hope to have the remaining 100 completed by the end of 2012.

Second, making data accessible. We've always been keen to make all our buildings' (including schools) energy use and costs available to as many people as possible. As a result, we now have a website where users can log on and see this information for all our buildings. And with the introduction of smart metering, they can now see the energy consumption in buildings on a daily basis.

What do you love about your job?

It's very satisfying when you reduce a building's energy consumption and expenditure and improve its cash flow. Working for a Local Authority makes this extra special as the money can then be spent to improve alternative services valuable to the public.

What do you think are the biggest challenges?

Our challenge is to limit the financial impact related to energy usage across the local authority. This will be done through sensible energy procurement while continuing our efforts to reduce the energy consumption across our building estate.

What are you most proud of?

I recently finished an energy project improving the lighting at three of our primary schools. I completed the energy surveys, recommended the type of replacement lights and analysed the results. We improved the lighting levels, reduced the electricity consumption, improved the energy awareness throughout the schools, and improved the schools' display energy certificate ratings.

What's the best way to engage employees on energy efficiency?

It's important to adjust your message depending on who you're talking to. For example, when I talk to finance managers, I only ever talk about the financial implications of reducing energy consumption. However, when talking to environmental officers, I only ever talk about CO_2 levels.

9
Renewable electricity

Thus far, we have examined a comprehensive range of possible methods to bring down demands for energy in a building. Now we turn to low carbon electricity generation, either to satisfy some or all of the remaining demand, or to sell on. If the reader has skipped ahead, they may have found, in the Conclusion, evidence for why sometimes installing renewable electricity-generating equipment is not the most carbon-effective of options. Nevertheless, for various reasons, such as providing an income stream, it can be very attractive for a company to install their own renewable electricity generation.

Renewable energy comes from natural resources such as sunlight, wind, rain, tides, waves and geothermal heat, which are renewable and sustainable because they are naturally replenished at a constant rate. Their use displaces fossil fuels and helps curb the increase of the concentration of greenhouse gases in the atmosphere. The main options available for electricity are as follows:

- hydroelectric power;
- renewable energy CHP;
- wind power;
- solar photovoltaics;
- solar CSP;
- hydrogen fuel cells and other forms of storage.

Recent research by utility company EDF Energy estimates that the renewable energy generation market in the UK will double over the next five years and reach 75 TW-hours. That's the equivalent of about 20 per cent of the country's total generation capacity. A similar trend is being experienced throughout the world. The reason is simply that the cost of the technology is coming down. The main drivers for the market are expected to be independent generators, principally in the wind and biomass sectors. This is an area of significant growth, especially driven by those organisations and companies, including local authorities, in the waste and water management business, utilities and industrial or manufacturing units. The report says that by 2017 this may account for over 60 per cent of the total renewable generation market. Existing independent specialist generators are expanding their operations, and many businesses are installing their own independent generation operations, often using long-term power purchase agreements with specialist suppliers who obtain finance and benefit from the public subsidies. If they want to capture the full benefit of their investments, EDF advises businesses to use a contractual arrangement of this type to help them manage the risk. It would combine the electricity they purchase from the grid and self-generated electricity into a single contract with the utility that helps with maintenance, and manages their exposure to wholesale electricity market price fluctuations. Of course, EDF would say that, and a confident organisation could also tread its own path, with the prospect of even greater potential financial advantages.

General points

Before moving on to discuss the individual technologies, it is worth noticing that the 'balance of system costs', which is to say everything except the generation plant itself, and which includes control technology and connections to the network, form a considerable part of the total cost; in some cases more than half. There are therefore advantages in scale, which make ambitious installations more financially worthwhile. This is true of all the technologies. Most of them are modular: one can always add more anaerobic digestion chambers, wind turbines, photovoltaic modules or solar collectors providing there is space on a suitable site. Systems may be expanded later, following experience gained or greater budget.

Renewable generators have little impact on the environment, but some people may object to some installations on aesthetic grounds. Planning permission will be required in most cases. Installation requires negotiation with the utility or local electricity grid managers, typically handled by the installer. Intermittency is an issue with all except hydroelectricity and biogas/biomass burnt in a CHP plant. Consequently, either backup or storage may be required, although usually the grid is used as a backup, with the advantage that surplus electricity not required at the time of generation may be exported for sale. Most (99%) renewable energy systems are grid-connected. It is best to commission experts to design a system; it almost goes without saying that one should tender for competing quotes.

Costs

Almost all renewable energy technology has a higher up-front capital cost than investment in equivalent-capacity fossil fuel generation plant. However, on a life cycle, levelised cost basis, it is often cheaper, because there is no fuel to purchase and lower operating expense. It is the only form of power generation that is reducing in price; fossil fuel prices are turbulent and rising. Therefore installing renewable electricity improves energy security (with storage backup in particular) and gives confidence on energy prices for decades to come. It also provides an income stream for any energy that is sold on or which receives incentives.

Whether a project makes any financial sense will depend on the metering arrangements (Feed-in-Tariffs (FiTs) in the EU, Renewables Obligation Certificates in the UK, and net-metering in the USA generally, but there are many exotic variations, too numerous to detail here) so a really good understanding of local metering regimes is necessary to evaluate any proposal. Financing packages are available from some specialist financial institutions and energy service companies that repay the cost of installation from the sale of electricity generated. Installations are often eligible for production tax credits, too, depending on their

Table 9.1 The average (based on weather data) 2011 levelised cost of electricity in the UK, without subsidies, given a 10 per cent discount rate, at projected Energy Procurement Contract prices, first-of-a-kind/nth-of-a-kind mix. Levelized costs are calculated by dividing the discounted sum of lifetime costs by the amount of energy produced to give a £/MWh figure. The costs of onshore wind, and especially PV, have come down significantly since the time these figures were arrived at

Technology	Levelised cost £/MWh
Energy from waste	−30
Landfill gas	44
Hydro 5–16MW	73
Co-firing conventional	74
Sewage gas	83
Onshore wind >5MW	91
Anaerobic digestion <5MW	105
Biomass 5–50MW	8
Offshore wind > 100MW	31
Hydro <5MW	131
Biomass CHP	134
Bioliquids, biodiesel	306
Solar PV >50kW	314

Source: Department of Energy and Climate Change, 'Review of the Generation Costs and Deployment Potential of Renewable Electricity Technologies in the UK', June 2011

location. There are sometimes advantages to generating power close to the point of use, as losses in transmission are reduced.

Estimating cost saving

To make the financial case, the cost saving associated with the generated electricity needs to be worked out. This depends on whether it is used directly or exported. Electricity used directly is valued at the unit cost for purchased electricity. Electricity exported is valued at the price for electricity sold to the grid. The effective price depends on how much is used directly. This factor (F) is in the range 0.0 to 1.0 and its value depends on the coincidence of electricity generation and electricity use. The fuel price to calculate the cost benefit is then:

F × normal electricity price + (1 − F) × exported electricity price

For example, if 75 per cent of electricity is used on-site and 25 per cent exported, and the price of imported electricity is 12p/kWh, while the price fetched for electricity sold is 8p/kWh, then the cost benefit of using it on-site is:

$$(.75 \times 12) + (.25 \times 8) = 9 + 2 = 11p/kWh$$

Estimating CO_2 emissions saved

To calculate the CO_2 emissions saved, the emissions factor is that for the grid-displaced electricity, which will depend on the location of the installation and the make-up of the grid-supplied electricity (see the Appendix for sources). The same factor applies to all electricity generated, whether used within the building or exported. The total number of kWh generated in a year would be multiplied by this factor to determine the number of kilograms of carbon dioxide saved.

Choosing a technology

The first stage is to survey, and if possible quantify, which renewable technology is appropriate for the site. It is no good thinking that a photovoltaic, roof-mounted system is desirable when, for the sake of argument, a wind power system would be more appropriate. When subsequently talking to a solar installation company, they would not point this out. Questions to ask would include the following:

- Is there a stream or river nearby that could be used for a hydroelectric scheme?
- Is there an accessible, Equator-facing site which is not shaded?
- Is the site windy for much of the year? Is there no turbulence nearby?
- Is there a ready source of fuel for biomass, anaerobic digestion, or energy from waste plant?
- Would it be possible to use the heat as well as electricity generated in a CHP plant?
- Is there a source of waste oil?
- Is there a source of landfill or sewage gas?
- Is there an existing coal-fired generator which could be used for co-firing with biomass?

Capacity factor

It is also worth thinking about the capacity factor of the technology. The capacity factor of a generator is the ratio of its actual output over a period of time compared to its potential output if it had operated at full nameplate capacity the entire time. It is calculated by dividing the total amount of energy the plant produced during a period of time by the amount of energy the plant would have produced at full capacity.

Intermittent renewable energy sources like solar power and wind power have low capacity factors, typically less than 25 per cent. On the other hand, hydro-electric schemes and generators relying on a readily available fuel (anaerobic digestion and all plants which burn fuel) have a higher capacity factor since they typically run most of the time, unless deliberately idled for maintenance and so on.

Sourcing and talking to suppliers

Whatever system is to be installed, when sourcing suppliers it is vital to bear the following points in mind:

1 Always choose a supplier who has been in business for some time and obtain independent testimonials from previous clients.
2 Obtain at least three quotes from different suppliers.
3 Make sure the installers are members of the REAL Assurance Scheme (in the UK) (http://www.realassurance.org.uk) or the North American Board of Certified Energy Practitioners (NABCEP), whose members agree to abide by their Consumer Code.
4 Always be sure to compare like with like; the same number of components with comparable specifications.
5 Make sure the suppliers explain how they have calculated the size of the system to be appropriate for your needs, supply clear information and operating instructions, provide an estimate of how much electricity will be generated by the proposed system, and what this is as a proportion of your annual use.
6 Secure guarantees valid over the expected life of the system and insurance are essential.

The most efficient systems are often more expensive; quotes should not be compared on price alone, but on overall cost/benefit over the lifetime of the system (25 years or more).

General design advice

The following advice comes from experience of many systems:

- Keep it simple: increased complexity reduces reliability and increases costs, especially for maintenance.
- Calculate the total life cycle costs and benefits, factoring in the capacity factor, to compare potential systems with alternatives; be realistic when calculating the capacity factor, i.e. how much of the time it will actually be generating.
- Plan for periodic maintenance: renewable energy systems have a good reputation for unsupervised operation but all require some degree of monitoring and care.
- Be realistic when estimating loads: including a large safety factor can increase costs substantially.
- Repeatedly check weather data in the case of wind and solar power: errors in estimating the resource can cause disappointment.
- Different hardware with different characteristics have different costs. They may also be less compatible. Investigate thoroughly all options before deciding on the optimum combination.
- Ensure the system is installed carefully: each connection must be made to last 30 years, because it will do so if installed properly. Deploy the correct tools and techniques. System reliability is no higher than the weakest connection.
- Electricity is hazardous: be rigorous about safety precautions during installation and in operation. Comply with local and national building and electrical codes.

Hydroelectric power

Hydroelectric power requires a consistent flow of water together with a drop in level that is sufficient to provide the force required to turn a turbine. This makes the technology very site-specific, but for those situations where it is appropriate it is highly reliable and cost-effective. The technology is over 150 years old, and well proven. A system, once installed, can last for up to 100 years, so it is more cost-effective in this sense as well.

The greater the volume of water flow, and the greater the head, or vertical distance it drops, the more electricity will be generated. The more power that can be generated, the more cost-effective it is. A small flow will not produce much power. Therefore the power in the stream must be measured first, as part of a feasibility study.

To do this, for small streams the flow is diverted into a receptacle of known volume, and the time taken to fill it measured. From this, a flow rate in litres per second, or gallons per minute can be calculated. To get a rough idea in the case of larger watercourses, the area of a cross section is calculated using the width and the depth, then the average speed between two points 10 metres or yards apart is measured. From this the flow rate can be deduced. A more accurate idea is obtained by using hydrological maps and flowmeters or stream-gauges, which take readings of the flow at 15- or 30-minute time intervals. Measurements are taken at different times of the year to allow for any dry spells.

From the flow rate, the expected electrical output of the turbine may be calculated by using the following formula:

Power out (in kilowatts) = the flow (Q) × the head (in yards or metres) (H) × the specific weight of water (Y, or 9.81 kilonewtons/m^2)

For example, a scheme with a flow rate of 0.5m^2/second over a 12-metre head would in theory produce an output of 0.5 × 12 × 9.81 = 58.86kW. In reality, efficiency losses usually account for around half of this output, making it more like just under 30kW. Multiplying this by the length of time in hours over which the generator would be expected to operate in a given year or month will give you the number of kilowatt-hours for that period. Different turbines are efficient for different heads and flow rates, and specialist help should be sought in choosing the right type.

Planning permission must be sought and a licence obtained from the Federal Energy Regulatory Commission (FERC) in the USA, the Environment Agency (in England and Wales), SEPA (in Scotland) or Environment and Heritage Service (in Northern Ireland). The Environment Agency estimates that there could be around 1200 licences granted for hydropower by 2020. In the USA it is also recommended to consult the US Army Corps of Engineers and the US Fish and Wildlife Service.

The components of a micro-hydropower system are as follows:

- A water conveyance, which is a channel, leat, pipeline, or pressurized pipeline (penstock) that delivers the water.
- A turbine, pump, or waterwheel, which transforms the energy of flowing water into rotational energy.

- An alternator or generator, which transforms the rotational energy into electricity.
- A regulator, which controls the generator.

There is usually an inverter to convert the low-voltage direct current (DC) electricity produced by the system into 120 or 240V of alternating current (AC) electricity. Optionally, a dam may be constructed as a holding bay for the water for times when rainfall is low, but these are very unusual in small, micro-hydro projects.

Case study: Osbaston hydro scheme, Wales, UK

This 0.15MW hydropower installation, located in Osbaston, Monmouthshire, Wales, and opened in 2009, uses an Archimedes screw type of turbine, which is thought to be more fish friendly than some other designs. It produces around 670,000kWh of electricity per year, sufficient to power 152 homes. Waste heat is supplied to a nearby home. A fish pass is built alongside it, allowing salmon to navigate the river. The scheme replaced an earlier one which had been on the site since 1896.

Figure 9.1 An Archimedes screw hydroelectric installation at Osbaston, Monmouthshire, Wales.

Source: Environment Agency

Combined heat and power

Renewable electricity from combined heat and power comes from two sources: the combustion of renewable gas (biogas), and the combustion of biomass. Plant sizes tend to range from around 1MW to many MW of electrical generation capacity.

Biogas

Biogas can be produced from anaerobic digestion or sewage gas. Since the latter would be an unusual source in most locations, we will move on to anaerobic digestion. This topic is covered from the heat and fuel source perspective in Chapter 7.

Figure 9.2 Gas storage tanks for an anaerobic digester, running on farm waste.

Source: Author

Figure 9.3 Biogas CHP units. These models are capable of producing up to 400kW of electrical energy and 550kW of thermal energy, making them ideal for apartment blocks, hotels, commercial and industrial buildings.

Source: Viessemann

The advantage of biogas is that it can be stored and burned when required. Before being used it must be cleaned of hydrogen sulphide because of its capacity to corrode metals. Used in a CHP, or cogeneration, plant, the gas is fed to an engine, where it is burnt to transform water into steam, which turns a generator to create electricity. The waste heat is harnessed, and some of it is used to heat the anaerobic digester, which needs heat to work. Up to 45 per cent of the energy in the gas can be converted into electricity. Roughly, this means that one cubic metre of methane, with an energy potential of 10kWh, can produce 4.5kWh of electricity. The thermal aspect of the system is about 40 per cent efficient. This means that he CHP unit altogether can utilise about 85 per cent of the energy in the gas.

Biogas may also be used in micro-gas turbines (micro-CHP), although the electrical efficiency is generally lower (around 28 per cent), and units are still fairly expensive. It may also be used as a feedstock for a fuel cell, where electrical efficiencies can reach up to 50 per cent. Finally, a further application is absorption refrigeration for cooling processes, a topic also dealt with in Chapter 7.

Figure 9.4 A fuel cell designed to work on biogas.

Source: Planetary Engineering Group

Biomass fuels

Some biomass fuels are more suitable for the small scale, while others are scalable. Table 9.2 summarises these applications. They also vary according to their calorific value, their moisture content and their energy density, as well as the volume which they take up, all important factors in designing a plant. Table 9.3 gives average values for these properties. A plant is usually installed following discussions with an energy services company (ESCo) who can help identify a local source of suitable fuel and provide specialist help in developing its appropriate use.

Table 9.2 Biomass fuel types for CHP

Fuel	System size	Operation
Woodworking offcuts/sawdust	<50kW$_{th}$	Daily, manual input
Common agricultural commodities such as spent grain	<50kW$_{th}$	Daily, manual input
Logs	<50kW$_{th}$	Daily, manual input
Pellets	<150kW$_{th}$	Automatic feed
Chipped/shredded wood	50kW$_{th}$ – multi-MW$_{th}$	Automatic feed
Bales (straw or miscanthus)	<300kW$_{th}$	Daily, manual input
	multi-MW$_{th}$	Automatic feed
Municipal waste	multi-MW$_{th}$	Automatic feed

Figure 9.5 Wood pellets for use in a biomass boiler or CHP plant.

Source: The Carbon Trust

Table 9.3 Typical bulk, calorific and energy densities of different biomass and fossil fuels

Fuel	Net CV1 MJ/kg	CV kWh/kg	Bulk density kg/m³		Energy density by volume MJ/m³		Energy density by volume kWh/m³	
			Lower	Upper	Lower	Upper	Lower	Upper
Woodchips @ 30%	12.5	3.5	200	250	2500	3125	694	868
Log wood (stacked – air dried: 20%MC)	14.6	4.1	350	500	5110	7300	1419	2028
Wood – solid oven dried	18.6	5.2	400	600	7440	11,160	2067	3100
Wood pellets	17	4.7	600	700	10,200	11,900	2833	3306
Miscanthus (bale – 25%MC)	12.1	3.4	140	180	1694	2178	471	605
Anthracite	32.1	8.9	1100	35,310	9808			
Oil	41.5	11.5	865		35,898		9972	
Natural gas	–	–	–		36		10.13	
LPG	46.9	13.0	500		23,472		6520	

Source: Gastec at CRE Ltd and Annex A, Digest of UK Energy Statistics 2007

Burning wood is not without environmental cost. Not only does it produce carbon dioxide, which will not be replaced for 20 years if trees are replanted, but it also emits dioxins and persistent organic pollutants such as PAHs and PCBs, as well as particulates such as DM10, all of which can kill. The level of these pollutants is at its highest just after the burner has been lit, before it approaches its optimum operating temperature of 200°C (392°F). Even at this point, it only reduces the pollutants by about 50 per cent. Incinerators are legally impelled to operate at 800°C (1472°F), more effectively to break down these pollutants. Unseasoned wood produces even more pollution. Seasoning takes at least a year and requires a well-ventilated, dry environment.

It is vital to design the system suitably well for the delivery, storage and loading of the biomass fuel. Whatever solution is arrived at must be durable for the lifetime of the plant, and adhere to environmental standards to prevent pollution, for instance, by dust. Storage must allow for the fuel to dry, but not get wet when it rains. Fuel types may be combined in a hybrid system, as exemplified by the case study below.

Case study: Foresterhill Energy Centre, Aberdeen, Scotland

The £13.5 million Foresterhill Energy Centre consists of a 1.5MW biomass steam boiler fuelled with locally sourced woodchip, two 8.5MW and one 6.5MW dual fuel (gas/oil) steam boilers and a gas turbine (Centrax 501-KB7) combined heat and power unit, which produces 5.2MW of electricity plus 12 tonnes of steam per hour. Compared to meeting the energy demand of the site via the original supply

Figure 9.6 The Foresterhill Energy Centre, Scotland.

Source: Keppie Design

arrangements, the new centre will reduce CO_2 emissions by 9830 tonnes (16%) and energy costs by £2.95 million (39%) per year.

It is located at NHS Grampian's new health campus in Aberdeen, UK, and has been awarded a BREEAM (Building Research Establishment Environmental Assessment Method) Excellent rating score of 83.32 per cent, as well as being named best industrial building at the BREEAM Awards 2012. Biomass storage is conventionally achieved via a walking floor arrangement linked to either a ram or chain grate stoker. An integrated fuel storage and delivery system reduces the overall footprint of the building by 450m², and eliminates dispersion of dust to the atmosphere. The design was fully developed using building information modelling (BIM). The walk-through visualisation generated from the BIM model was instrumental in validating the adequacy of the access and maintenance arrangements.

With biomass plants, correct sizing is everything, and this is related more to the heat output than to the electricity output, since the latter is used all the year round. Biomass CHP plants are best operated relatively continuously at between 30 per cent and 100 per cent of their rated output. This means that systems supplying buildings that are generally only occupied during working hours have the lowest capacity factor, because they only need heating during working hours and in winter months. Conversely, 'service' applications such as swimming pools and hospitals, and process applications, such as horticulture and food and drink manufacturing, will have a higher capacity factor, since they are used more often and continuously. They therefore tend to be significantly more cost-effective.

Table 9.4 Typical capacity factors for different biomass CHP applications

Category	Typical capacity factor
General occupancy building	0.2 (20%)
Service applications	0.45 (45%)
Process applications	**0.6 (60%)**

Wind power

In general, wind turbines only have a chance of working at maximum efficiency in open, exposed spaces. Lower wind speeds in urban areas, coupled with the turbulence caused by nearby buildings or trees, mean that the wind speed of 5m/s (11mph) required for contemporary turbines to operate efficiently is rarely reached in an urban situation.

This has been established by a substantial monitoring exercise of real micro-wind turbines both on urban rooftops and free standing in rural areas[1] published in July 2009 by the UK EST (2009). To operate efficiently, turbines do require a tower and no surrounding obstructions in the direction of the prevailing wind. As the report says: 'a properly sited and positioned 6kW rated free standing pole mounted turbine . . . would be expected to generate approximately 18,000kWh per annum'. This represents a very quick payback.

The performance of free-standing turbines in the survey above frequently exceeded the manufacturers' quoted annual load factors of 17 per cent; the average was 19 per cent and the best were 30 per cent. By contrast, 'No urban or suburban building mounted sites generated more than 200kWh per annum, corresponding to load factors of 3% or less'.

Turbines come in many sizes and power ratings, from 3kW upward. The largest, offshore, machines have blades the length of a football pitch, stand 20 storeys high, and have an output sufficient to power 1400 homes. A 5–15kW domestic machine has rotors 8 to 25 feet in diameter, would be 30 feet (10 metres) tall and could supply the needs of a small business. Utility-scale turbines range from 50kW to 1MW.

Table 9.5 The classification system for wind turbines

Scale	Rotor diameter	Power rating
Micro	< 3m	50W–2kW
Small	3m–12m	2kW–40kW
Medium	12m–45m	40kW–999kW
Large	> 45m	> 1999kW

In general, small turbines operate at variable speeds, and output direct current (DC) electricity. Larger ones are geared to operate at a constant speed and alternating current output (AC) electricity. A typical small wind generator has a rotor that is directly coupled to the generator, which produces electricity either at variable frequency 120/240 volt alternating current, or at 12/24 volt direct current for battery charging. Control equipment for machines that generate DC (the smaller machines), inverters that convert the DC electricity to mains quality AC are necessary. Larger turbines produce three-phase AC. Small turbines are often used in combination with photovoltaic modules to power off-grid applications, from street furniture to telecommunications systems. They make a good pairing because the sun is often shining when there is no wind, and vice versa. They are supplied as a turnkey kit complete with battery storage for up to three days' backup.

Wind resources are characterised by wind power density classes, ranging from class 1 (the lowest) to class 7 (the highest). Good wind resources (class 3 and above, which have an average annual wind speed of at least 6m/s (13mph) are found in many locations. But turbines do need an uninterrupted wind flow, and the higher they are above the ground the more wind there is. This is why they are commonly mounted on poles 10m or 25m above the ground and in exposed places.

Calculating the power

Figure 9.7 An anemometer being used to measure the wind speed.

Source: Wikimedia Commons (NOAA Photo Library, NOAA Central Library; OAR/ERL/National Severe Storms Laboratory (NSSL))

The annual average wind speed needs to be greater than 5m/s (11–13mph). Wind resource maps are available on the National Renewable Energy Laboratory website, and the Google Application Google Earth, or, in the UK, the RENSmart website (www.rensmart.com/Weather/BERR). Having established that it is worth investigating the wind resource at your location, the site of the planned wind turbine should be monitored with an anemometer for 12 months.

Once the average wind speed is known, it is useful to examine manufacturers' brochures. These contain power/energy curves like the one illustrated in Figures 9.8 and 9.9, from which can be read off the level of power it may be possible to generate from a given wind speed.

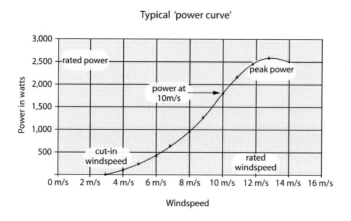

Typical 'power curve'

Figure 9.8 A power curve for a 2.5 kW turbine with a cut-in speed of 3 m/s, showing how much energy it will generate at different wind speeds according to its rating.

Source: Centre for Alternative Technology, Wales

The larger the diameter of the turbine, the greater the swept area and the more power is collected. This helps to explain how the power output increases with the cube of the wind speed, so a doubling of wind speed would result in eight times more power. A wind turbine will only operate at its maximum efficiency for a fraction of the time it is running, due to variations in wind speed. In other words, the longer the high wind speeds last, the more energy will be generated.

But no wind turbine can convert more than 59.3 per cent of the kinetic energy of the wind into mechanical energy turning a rotor. This is known as Betz' Law after Albert Betz, the German physicist who discovered it. The proportion that can be converted is the 'power coefficient' (C_p) of a turbine. In reality, wind turbines cannot operate at this limit. The C_p value is unique to each turbine type and is a function of wind speed within which the turbine is operating, and is inside the range 0.35 to 0.45 even in the best-designed machines. Once the other

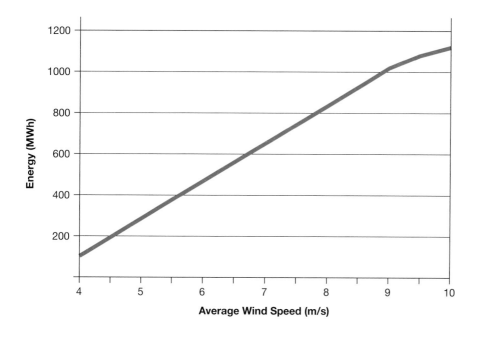

Figure 9.9 A graph showing, for different wind speeds, the expected output of a wind turbine. Manufacturers often supply graphs of this nature for each wind turbine they produce.

Source: Author

factors in a wind power system, such as the gearbox, bearings and generator, are taken into account, only 10 to 30 per cent of the power of the wind is ever actually converted into usable electricity.

There are various important wind speeds to consider:

- Start-up wind speed: that will turn an unloaded rotor;
- Cut-in wind speed: at which the rotor can be loaded;
- Rated wind speed: at which the machine is designed to run (the optimum tip-speed ratio);
- Furling wind speed: at which the machine will be turned out of the wind to prevent damage;
- Maximum design wind speed.

A careful matching of the electrical energy requirements should be made to maximise the use of the wind power.

Solar power

Like wind, solar power is site-specific. The relevant solar technologies are as follows:

- photovoltaics (PV) (using the sun's light);
- concentrating solar thermal power (CSP) (using its heat).

Photovoltaics are by far the most versatile and lowest cost form of solar power. In some parts of the world it can now, under certain conditions, compete with grid electricity even without a subsidy. Solar PV has the following advantages:

- Low maintenance, due to no moving parts;
- Durability: modules may continue to provide free energy for up to 30 years, yielding a feed-in-tariff income stream for eligible grid-fed electricity, although electrical output can tail off slightly;
- Scalability: by adding more modules together, the amount of power that can be generated is only limited by the available space and budget;
- Quick to install: small solar power stations can be erected in two months, a medium-sized roof-mounted system in one week.

But there are disadvantages: besides being intermittent, working only when the sun is shining, it is seasonal, particularly in northern latitudes, so when electricity is needed most, there may not be enough available. Therefore PV systems must be appropriately sized, or supplemented with other sources of electricity. In 99 per cent of cases this will be the grid. Modules output also declines, though very gradually: they may produce 95 per cent of rated output for the first ten years, 90 per cent for the next ten years, etc. The rate of deterioration depends on the type of cell used. The site must have no shading falling upon it at any time of day.

Developers or contractors conduct site surveys using instruments which measure the amount of solar irradiation received at that location, broken down

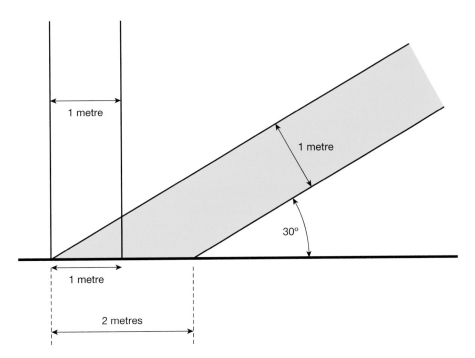

Figure 9.10 The angle at which the light hits a silicon module is important. Ideally, it needs to do so straight on. At an angle of 30 degrees, for example, the same amount of light will be spread over double the area of the module, halving its effectiveness.

Source: Author

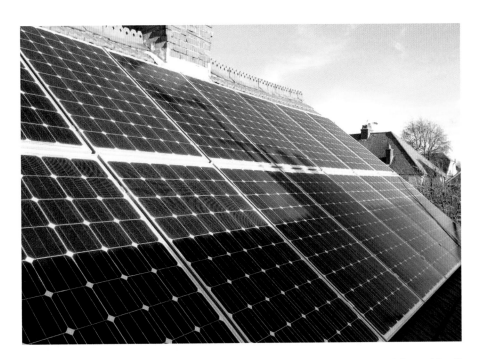

Figure 9.11 An example of poor siting: there needs to be no shading on the module, as this will prevent it from generating electricity. Even a small amount of shade can reduce output disproportionately.

Source: Author

Table 9.6 The effect of tilt and orientation on a solar module's performance at latitude 52° (Birmingham, UK). The table shows that the efficiency of a PV array is compromised by 17 per cent when placed vertically on a building façade, and that east- or west-facing arrays suffer a 50 per cent loss

Orientation	Tilt	Generation (kWh/m²/day)	Difference from optimum
South	35°	3.00	Optimum (%)
South	Vertical	2.18	17
South	Horizontal	2.63	12
East/West	35°	2.46	18
East/West	Vertical	1.63	47
East/West	Horizontal	2.63	12

Source: BRE

into several components such as direct, diffuse and reflected light, and feed them into functions that take into account the above variables plus the azimuth. They compare these to both freely and commercially available figures.

The amount of power produced by a solar cell depends on its efficiency and how much light is hitting it. A c-Si (silicon) cell will produce most power when pointed directly at the overhead summer sun on a bright, clear day. To enable them to be compared, modules are characterised by 'peak power' (watts-peak or Wp) output. It does not mean that they will always produce this amount of power. Peak power, as measured under internationally agreed Standard Test Conditions (STC), is what is produced when they are exposed to 1kW per square metre of light at 25°C (77°F) in an atmospheric air mass of 1.5.

Typically, modules are roof-mounted, but in some cases they are integrated into the building structure. They are usually grid connected. Larger solar farms are being increasingly constructed, because they are cheaper per kilowatt of output, and they take advantage of government support for larger projects. Southern areas of the UK are seeing solar parks of between 2 and 5MW being constructed where there is a large area of flat or south-facing land of agricultural grade 2b or lower covering between one and four fields, with minimal shading. Old aerodromes and farmland are proving popular. The site should also be close to a 33kV electricity substation and be out of sight of residential properties in order to receive planning permission. A 25-year lease from the developer would be part of the contractual arrangement. Income could be around £1000 per annum per acre, and the contract should specify a cheap rate for energy consumed directly by the client.

In the UK, ROC (Renewable Obligation Certificate) payments have made solar parks attractive. Feed-in-tariffs are only eligible for solar parks with a maximum output of 5MW, which would typically be 25 to 30 acres, whereas ROCS have no limit. However, the ROC regime ends in 2018, to be replaced by the carbon price floor support scheme. In America, there are various federal tax incentives and 10 per cent tax credits, at least until 2017. Grid parity for unsubsidised solar energy is expected to happen in some states relatively soon, particularly in New York state and California. Some seven states are also very attractive, according to the Institute for Local Self-Reliance. In central Europe, the levelised cost of c-Si PV over 25 years is currently estimated at 0.2 euro/kWh (200 euro/MWh), according to researchers at Imperial College, London.

Case study: The Port of Workington, Cumbria, England

A 50kWp solar PV system was installed at the Port of Workington, Cumbria, England, in 2012 as part of the Cumbria County Council-owned port's ongoing £5.7m regeneration programme, jointly funded by Britain's Energy Coast, Nuclear Management Partners and the Nuclear Decommissioning Authority. Comprising 200 high-efficiency 250kWp solar PV panels, the system is expected to generate 38,440kWh of electricity in the first year.

Martin Cotterell, founder and technical director of the developer Sundog Energy, expects, given feed-in-tariff support, payback for a system like this in little over seven years and a projected internal rate of return of around 15 per cent. The wind turbines have been in place for a few years and export electricity directly into the grid.

System components

A typical system includes: the modules, a grid-inverter, which is a different type of inverter to that used in stand-alone (non-grid-connected) systems, controllers, disconnects, meters and fuses. A system that is also required to supply power when the grid is down would include a battery-connected inverter and relay controller.

Calculating output

The annual energy that can be produced by a given array of modules depends on several factors, most especially the peak kilowatt of rated power supplied by the manufacturer of the module and the insolation data for the location. Public sources of insolation at most locations are available from PVGIS or the US NREL. This will give the average amount of energy available per square metre for that location – in kWh/m² – for the whole year. To calculate the estimated output per year per PV module, use the following formula:

Module rated output (peak watts) × peak sunshine hours (per year) x 0.75 (performance ratio) = energy generated (kilowatt-hours/year)

For example, 1500W × 900 peak hours per year × 0.75 = 1,012,5005 watt-hours (Wh) or 1102 kilowatt-hours (kWh) (approximately)

Other factors to be taken into account include the following:

Figure 9.12a and b
Sample solar energy
insolation maps of
Europe and the USA.

Source: (a) PVGIS; (b) US
NREL

Photovoltaic Solar Electricity Potential in European Countries

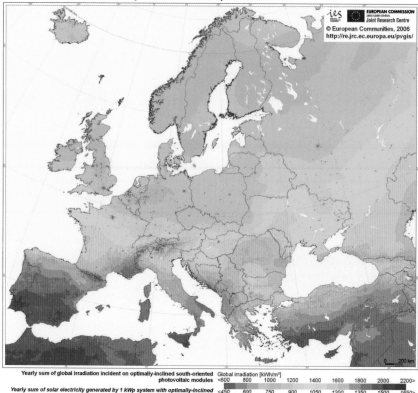

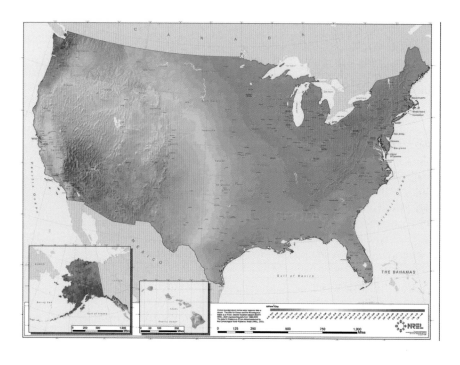

- The orientation (azimuth) (-90° is east, 0° is Equator-facing and 90° is west).
- The tilt angle of the modules from the horizontal plane.
- The energy conversion efficiency of the modules.
- The extent to which their efficiency is affected by temperature.
- Factors to do with the clarity of the atmosphere and the path of the sun.
- Possible shading at any time of day and year.
- System efficiencies, such as those of the chosen inverter and the wiring.
- The type of mounting structure: fixed or tracking (tracking increases output).

Case study: PV installation, UK

Figure 9.13 The solar array being installed on the roof of the Huntapac factory.

Source: Huntapac

A 50kW PV array was installed at the warehouse of vegetable producer Huntapac Produce in Tarleton, Lancashire, England. Estimated savings are put at £500,000 and 0.5 tonnes of CO_2 over the next 25 years. A series of ZN Shine 190-watt monocrystalline gallium panels were connected to a SMA Tripower 15000 TL G59-2 relay inverter. A challenge of the project related to keeping installation time as short as possible to minimise disruption. The power supply was connected within a two-hour time slot on a Sunday. The modules were fitted directly onto the roof and the inverters were situated in an existing roof void.

A year on from installation, the array has performed very much in line with the company's expectations in terms of energy output and the income generated via the feed-in-tariff. The work was carried out by Renewable Solutions UK Ltd, a company that is registered under the Microgeneration Certification Scheme, a

> consumer protection scheme, and is an agent for Carbon Trust Commercial Funding packages.
>
> The work was part of an ongoing programme of energy efficiency measures that include the use of an integrated utility management dashboard to manage the use of water, electricity and gas, the substitution of more energy-efficient lighting throughout the premises, and the use of variable-speed motors in the packaging plant. The building itself is essentially a standard shell but the building fabric will be improved in a later phase.

New technologies are becoming cheaper as well, including thin-film modules and dye-sensitised solar cells, but they are still a little way off from hitting the mainstream market. The latter have the advantage that the photo-sensitive dyes can be printed on many types of surface, including glass and steel for building exteriors. They do not need to be pointed directly at the sun. In the mid-term future we should begin to see buildings clad in this fashion, generating electricity.

Concentrated photovoltaics

Concentrated photovoltaics (CPV) use lenses like magnifying glasses to concentrate sunlight from a wider area onto a smaller area where the PV cell is located. This reduces the number of cells required, making an array cheaper, and reduces the amount of land required for the same amount of power output. They are suitable for solar power stations in the sunbelt areas all over the world, where normal direct irradiation is greater than $1800 kWh/m^2/year$.

Electrical energy storage

Energy storage systems enable electricity generated at a time of low demand to be stored and used at a later time when electricity demand is high. They go hand in hand with the development of the smart grid. The UK's Low Carbon Innovation Coordination Group estimates that, combined with the smart grid, storage solutions could save the UK between £4 and £19 billion in deployment costs up until 2050. In the USA, various projects have been funded by the American Recovery and Reinvestment Act (ARRA) with $185 million of public money, which attracted $585 million of private investment.

Steven Berberich, the President and CEO of the California Independent System Operator, is of the opinion that 'storage plus renewables is a marriage made in heaven'. Energy storage significantly increases the effectiveness of wind- and solar-generated electricity because the energy is time-shifted to peak demand, which strengthens the business case for investment in a renewable generation scheme and means fewer generation plants need to be constructed. There is another advantage: if storage is located near the point of use, this reduces the need to invest in power delivery infrastructure and reduces transmission losses.

There are many storage technologies in development, and the following are those which may (in some cases eventually) be appropriate for non-utility applications:

- Ceramic bipolar batteries, being supported already by the UK Technology Strategy Board for use with PVs.
- Compressed air energy storage (CAES) and liquid air, where the main challenge is to develop adiabatic (zero heat loss) compression to improve efficiency.
- Flywheels, which are achieving ever higher rotation speeds (e.g. hubless design).
- Hydrogen, generated from renewable energy, and used in conjunction with fuel cells.
- Lithium-based batteries, where developers are improving solid-state conductors and their longevity.
- Sodium-based batteries, where the challenge is to improve durability and electrolytes (including solid-state).
- Redox and hybrid flow batteries, where the need is to develop low-cost membranes and real-time impurity sensing.
- Super-capacitors, where the challenge is to improve high-voltage electrolytes.
- Thermal-to-electric storage, where the energy needs to be quicker to access and convert when required.

Berberich believes that frequency regulation is the first market opportunity for energy storage, because it is already economic. A ruling in the United States by the Federal Energy Regulatory Commission (FERC) that forces independent system operators to take into account the benefit of storage is making them see its cost-effectiveness when used for this purpose. Two pilot projects are establishing this near New York: one, run by Beacon Power, uses 20MW of flywheel power; the other, by AES, uses 8MW of lithium batteries. Another frequency regulation energy storage project in EastPenn, Pennsylvania uses 3MW of innovative batteries that look like lead acid but with one electrode containing carbon; a cross between ultra-capacitors and lead acid batteries with ten times the cycle life of other batteries.

Hydrogen storage involves an electrolyser splitting water into hydrogen and oxygen, powered by intermittent wind or solar power. The hydrogen may be used in hydrogen-fuelled vehicles or in static fuel cells to provide uninterruptible power supplies for remote or sensitive applications. This is not yet a cost-economic proposal for most buildings.

Note

1 Location, location, location: The Energy Saving Trust's Field Trial Report on Domestic Wind Turbines. EST, 2009, London.

Energy managers share their experience

Kit Oung, Energy Consultant

What does your job involve?

I'm on secondment to a large pharmaceutical manufacturing site where I work to reduce energy and water consumption. I have to identify opportunities to reduce process-based energy consumption and water consumption, and develop the business case for funding approval and implementation of the identified projects.

Before this I carried out energy audits either as paid consultancy work or as part of a business development activity to demonstrate a company's capability in delivering value for the potential client. Often I manage a small team of engineers studying opportunities to reduce energy consumption. I have managed several large industrial energy auditing projects worth £60k to £70k, and several Grade 1 and Grade 2 listed buildings.

In my own time, I represent the Institution of Chemical Engineers (IChemE) at the British Standards Institute (BSI) where I routinely review and comment on energy and environmental standards. I also chair the development of EN 16247-3 energy audit in processes standard and ISO/CD 50002 energy audit standard.

What qualifications did you gain?

I graduated from the University of Sheffield with B.Eng Chemical and Process Engineering with Fuel Technology. I then completed M.Sc.(Eng.) in Environmental and Energy Engineering. In 2010, I attended and passed the examination on British Standard Institutes' Lead Auditor course for BS EN 16001 Energy Management System. In 2011, I attended two short courses at Judge Business School, University of Cambridge: General Management Programme and the Professional Service Firm Leader. I am a Chartered Chemical Engineer, Chartered Energy Manager and a member of the Chartered Management Institute.

What does your day-to-day work look like?

- Discussing, sharing and formulating ideas and plans with the site's sustainability managers.
- Walking around the site, talking with the operators and technicians to identify energy reduction opportunities or carrying out the design of experiments to quantify the energy reduction based on data and writing up the business case.

- Assessing the risks to the business; determining if the work represents the highest return on investment for the client.
- When I'm satisfied that the proposal represents the best returns for the business, I'll write up the business case for capital funding. Once this has been approved, I act as the client representative.
- I then manage the projects and assess their effectiveness and success.

When I volunteer at BSI, I review, discuss and contribute towards developing the UK's position on energy and environmental systems standards and energy auditing standards. I attend presentations of new technologies and techniques from a wide range of industry sectors.

When I chair an energy auditing standards meeting, I act as a country-neutral person, to help develop a good-quality and usable standard. Through BSI, there are also opportunities to share my experience with others at conferences.

What do you love about your job?

Regardless of the science and approach to climate change, I genuinely believe reducing energy and water consumption is a win-win. I have been privileged to have carried out a variety of energy reduction work for a range of clients, including broadcasting companies, banks, listed buildings, pharmaceutical R&D, pharmaceutical manufacturing, food and beverage, milk and cheese manufacturing, distillery, petrochemical, chemical and specialty chemicals manufacturing. I enjoy meeting different people in different facilities and sites. I'm happy when my work generates an interest in the people I meet and in sowing the seeds for continued energy reduction efforts when I leave.

What do you think are the biggest challenges?

That the concept of 'energy management' has become very technical, data-focused and disconnected with day-to-day management activities. This manifests in several forms:

- Management thinks energy reduction can be achieved by purely installing the 'best' or 'greenest' technology with business-as-usual mentality. As such, energy reduction is treated as 'projects' that are compartmentalised into specific inputs, outputs and bounded timeframes.
- Managers tend to focus purely on energy reduction and ignore the other benefits associated with the energy reduction efforts (e.g. a reduction in quality defects, improvement in working conditions, reduction in maintenance requirements, release in capacity for additional production, etc.).
- Managers 'do' energy management for operational excellence reasons rather than for strategic and business competition ones. They take the short-term view, requiring short-term returns.
- They don't conduct employee engagement campaigns on energy-efficient behaviour, because they are difficult to quantify and therefore justify. Frequently, they expect an 'Energy Manager' to achieve all of the work without support and action from the whole organisation.

- Many of the technologies are not installed correctly or operated as intended. Some of the most frequent are (1) condensing boilers being operated in temperature ranges that do not allow condensation of flue gases, (2) variable-speed drives operated at 50Hz with large bypass lines, (3) heat pumps or CHPs are installed, yet the HVAC have excessive air turn-downs. There are a lot of other examples.
- Organisations engage on multiple similar-themed organisational programmes, thus confusing employees in a poor focus on energy reduction (e.g. simultaneously implementing ISO 9001, ISO 14001, ISO 50001, OHSAS 18001, 5S, Kaizen, TQM, TPM, and Lean Manufacturing).
- When implementing ISO 50001, many managers implement it as a separate management system, independent of the normal ways of working, and apply in a mechanistic way the requirements of the standard without understanding the principles. This generates an additional strain on the organisation.

These all stem from a very technical data management which is disconnected from the people side. They are issues I see in almost every organisation.

What are you most proud of?

I operate on the basis of: the clients are experts in their own business. I'm merely an external resource supporting the client to identify and support its implementation. The value of me to my client is that I bring my experience from past projects and what I've seen elsewhere. If necessary, I prefer to tell the client I don't know all the answers but I'm willing to explore them together with the client, rather than pretend to be the expert.

Some of my best energy, carbon and/or water reduction work involves a lot of client interactions and inputs. Employees from the client feel valued and involved. Frequently, they become the champions for the proposal to their management.

Disarming organisations' ingrained practices is important for the long-term sustainability of the business organisation, the environment, and future generations. I have written two books on energy management aimed at managers and busy business folks – trying to bridge the knowledge and re-engage interest by managers with energy. These are: *Energy Management in Business: The Manager's Guide to Maximising and Sustaining Energy Reduction*; and another book, co-authored with Graham Wooding, to help organisations implement ISO 50001 energy management system.

What's the best way to engage people on energy efficiency?

I've been asked by prospective students interested in becoming energy managers which subject to study at university. I told them:

- If you enjoy science and technology, study engineering and take up some electives or short courses such as organisation dynamics, microeconomics, behavioural economics, and management science.

- If you enjoy management subjects, study management and take up some electives or short sources on energy management, energy auditing, etc.

This is because it is not possible to manage energy purely by data. It is also not possible to manage energy purely by engaging employees without data. Both are required. This leads me to the next question: How do I engage my top management to become interested in energy management? Each organisation is different. Things that motivate someone to take action are different from person to person. I believe, and I frequently put it in practice, that it's best to use insider knowledge from within the organisation and the people that I meet. A balance has to be struck between the data I present and the needs of the audience. So, instead of looking for a one-size-fits-all communication and/or engagement tools, I advocate a 'use your knowledge about the audience, balance the technical and human content to meet the needs of the audience' approach. If you are able to, simplify the technical contents.

What is the first thing that you would recommend to save energy and carbon emissions?

When implementing a management system for energy:

1. Use a set consistent language within the organisation.
2. Involve the whole organisation.
3. Allocate responsibilities and resources early.
4. Challenge all established assumptions within the business.
5. Integrate the management system and ways of working into daily operations.
6. Utilise a set of appropriate performance measurements to the right level of responsibility.
7. Consider energy, carbon and water in the whole life cycle of the business and its value chain.
8. Set stretched but achievable objectives and targets.
9. Create opportunities to generate opportunities for quick wins.

When finding and implementing opportunities to reduce energy, carbon and/or water, instead of going out to buy the latest technology, first consider maximising the potential for reduction by:

1. Applying good maintenance.
2. Sizing the utilities and energy systems to match demand.
3. Make sure all equipment and machinery are installed correctly.
4. Turn off when not required.
5. Otherwise, operate the utilities and energy systems to match demand.
6. Control and minimise start-up and shut-down.
7. Apply good insulation.
8. Consider heat recovery.

What is your favourite tool of your trade?

I do not have specific or favourite tools I use with my clients. I use a range of low-tech tools and calculations highlighted in answers to the last question. For me, the most important tool kit I have is to:

- Understand the fundamental principles of energy management.
- Choose and use the right tools with insight.
- Focus on being competent on the fundamentals.
- Balance the hard science with people-based solutions.

10
Making the financial case

There are key barriers often found to the implementation of energy management measures, particularly in persuading senior management to make investments and commit to projects. These barriers may include a lack of understanding, conflicting priorities, misaligned financial incentives, hassle cost and lack of financial backing. Energy managers have to master successful tactics to overcome them. This chapter looks at several ways of presenting and comparing the financial case for different measures.

Marginal abatement cost curves

A marginal abatement cost curve (MACC) is a helpful, visual aid to providing an idea of the annual potential to reduce emissions and the average costs of doing so for a wide variety of technologies. We first met them in the Introduction. MACCs are a useful tool for cost-effectiveness analysis, but how are they compiled?

In Britain, the Committee on Climate Change (CCC)[1] has produced several MACCs for energy efficiency that incorporate research generated by three other important models:

1. BREDEM (the Building Research Establishment's Domestic Energy Model);
2. N-DEEM (the Non-Domestic buildings Energy and Emissions Model), which is based on detailed assessments of energy use in around 700 buildings, since they are extremely diverse in nature;
3. ENUSIM (the Industrial Energy End Use Simulation Model), originally designed to model industrial energy use by considering the take-up of energy-saving technologies in industry.

The MACC for the non-domestic sector is illustrated in Figure 10.1. The CCC concludes that, for the UK as a whole, there is:

a very significant contribution from improved energy management. These measures include turning monitors off at night, adjusting heating times or adding improved controls to lighting. These measures are almost entirely low cost measures with the potential to save over £800m countrywide per year for firms with very little (if any) up front expenditure. They could save over 8 MtCO$_2$ per year.

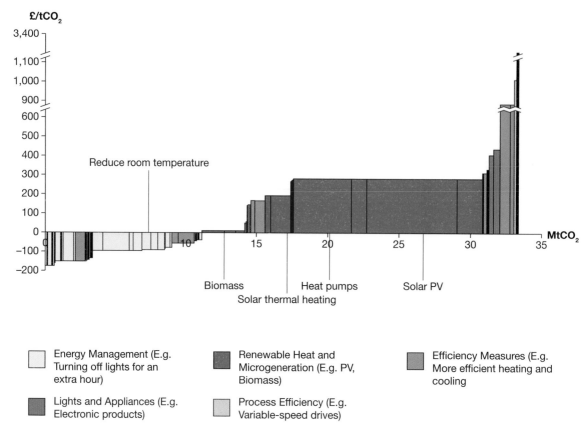

Figure 10.1 A marginal abatement cost curve (MACC) illustrating the technical potential for improvements in the non-domestic sector. Each column represents a particular measure. The vertical axis represents the cost per ton of carbon dioxide saved. The horizontal axis represents megatons of carbon dioxide saved throughout the lifetime of the measure. Measures to be taken on the left of the graph with columns descending beneath the horizontal axis have a negative cost; i.e. they save money. The ones on the right with columns ascending above the horizontal axis have a net cost; i.e. they cost more than they save. The further right that a measure is positioned, the greater its lifetime cost. All energy management measures have a negative cost and save money, as do many efficient heating and cooling methods.

Source: CCC

Estimating payback

MACCs are arrived at by calculating the payback for various measures. Projects are usually sold to management on the basis of return on investment. This may be expressed in two ways: as an effective interest rate based on the net present value; and as a payback period, i.e. the length of time it takes for the initial investment to be recouped by the savings earned or income generated.

Simple payback

The most basic of these is simple payback. However, it does not always illustrate the true benefits of an investment. Suppose an organisation demands a two-year payback period from any investment. Then, as the following example shows, it would miss out on the benefits of a project with a six-year payback period that actually had a better return on investment.

A project costing £60,000 which receives £30,000 in benefit per year following completion but which only lasts for three years would yield a total of £90,000. A project which costs the same amount, but only yields £22,000 per year, yet lasts for six years, would give a total of £132,000. However, if it were only evaluated on a two-year basis it would lose out to the three-year project.

A project which repays its cost every three years is demonstrably better than one which promises to return the investment in three years. To help establish this, the concept of discounted cash flow is introduced.

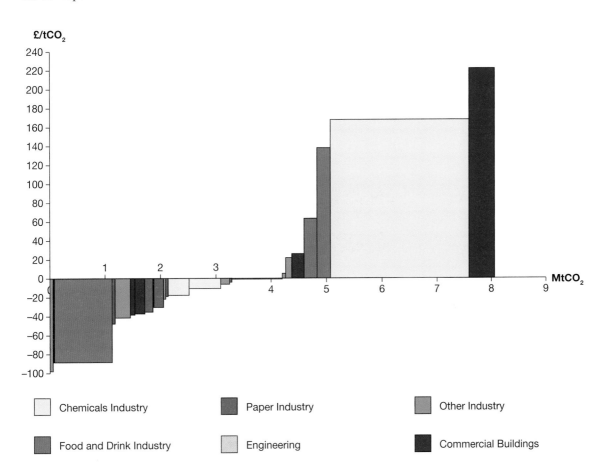

Figure 10.2 A marginal abatement cost curve (MACC) illustrating the potential for CHP (combined heat and power) in different sectors. It shows that even within a sector, whether a particular project is cost-effective depends on individual conditions. This is why, for each sector, there are different instances (illustrated by columns of the same colour), some of which are above the line (net cost) and some below the line (net benefit).

Source: CCC

Discounted cash flow (DCF)

Discounted cash flow provides a more realistic way of establishing payback. There are three stages for estimating DCF:

1   Estimate the resulting cash flow.
2   Apply the discount rate.
3   Calculate the end value (net present value).

The cash flow is taken from the estimated savings in energy cost resulting from the measure taken. This will depend on projections of future energy cost. For example, energy prices over the past three years may be projected on a median basis into the future, but this will then need to be discounted at a discount rate to be chosen. Discount rates are a function of the rate of inflation and represent what one unit of currency will be worth in a year's or 10 years' time. An average price [P] is calculated this way for each year of the projected lifetime [L] of the project. Each of these figures is then multiplied by the amount of energy [E] expected to be saved every year.

The lifetime period chosen for the project will depend on the expected lifetime of the technology. If it were a boiler, for example, it could be 15 years. Should it be an insulation measure, it could be 30 years. The total cost savings [S] generated by energy not used, compared to not doing the project, over the lifetime of the project will then be:

$$S = E \times [P_{(\text{year 1})}] + E \times [P_{(\text{year 2})}] + E \times [P_{(\text{year 3})}] \ldots E \times [P_{(\text{year L})}]$$

What discount rate should be chosen? The industrial model ENUSIM uses private fuel prices and a 10 per cent discount rate to reflect the incentives faced by firms. Some UK organisations adopt the rate used in the UK government Treasury's Green Book, which sets out the framework for the evaluation of all policies and projects, namely 3.5 per cent. Others simply adopt the current rate of inflation, or interest rate on a loan taken out for the purpose of the measure that would need to be repaid. It is useful to run the calculation several times with different discount rates.

Net present value (NPV)

The figure for the total cost savings [S] is not the final step in our calculation. We now need to deduct the cost [C] of taking the measure, which gives us a figure called the net present value [NPV] of the project. This is the value in today's money of all of the net profit that will be generated from adopting this measure. It is the most useful way of comparing the value of different measures. It takes account of the full value of the project and presents it in an easily comparable form. The net present value is therefore:

$$NPV = S - C$$

This is how all of the figures were arrived at that are represented in the MACC graphs above. Applying this to the two projects above, with a 10 per cent discount rate, lets us see the following:

Project 1 yields:

£30,000 (year 1) + £27,000 (year 2) + £24,300 (year 3) = £81,300, not £90,000

Project 2 yields:

£22,000 (year 1) + £19,800 (year 2) + £17,820 (year 3) + £16,038 (year 4) + 14,434.20 (year 5) + £12,990.78 (year 6) = £103,082.98, not £132,000

Both projects cost the same, £60,000. Subtracting this from the cost savings reveals that the NPV of the first is just £21,300, while that of the second is £43,082.98: over double that of the first project.

Internal rate of return (IRR)

The NPV can also let the projects be compared to what would happen to the same amount of money were it to be invested in a bank account with the same interest rate as the discount rate chosen. This is done by calculating the internal rate of return (IRR), or the interest rate on the investment, and is easily accomplished using Microsoft Excel as follows (together with Figure 10.3):

1 The initial expenditure is typed into a cell on a spreadsheet. This must be a negative number. Using our original example, -60,000 would be typed into the A1 cell.
2 The subsequent discounted cash return figures above for each year are entered into the cells directly under the first one. Following the example in Project 1, this would mean typing 30,000 into cell A2, 27,000 into cell A3, etc.
3 The IRR is then revealed by typing into the next cell beneath all the values the function command '=IRR(A1:A4)' and pressing the enter key. In this case, the IRR value, 18 per cent, is then displayed in that cell.

The IRR of the second project, calculated by the same method, is 20 per cent, and so provides a better rate of return. It is relatively easy to set up a template in Microsoft Excel to enable the performance of a similar calculation for any capital investment project. Further costs that are unique in any given year may be added, such as figures for additional maintenance, additions or repairs, and, at the end of the project, a figure for resale of any equipment, for example, its scrap value.

Presenting projects in such a way to senior management will allow them to compare their value with other projects they may be considering, as well as enabling the energy manager to prioritise projects.

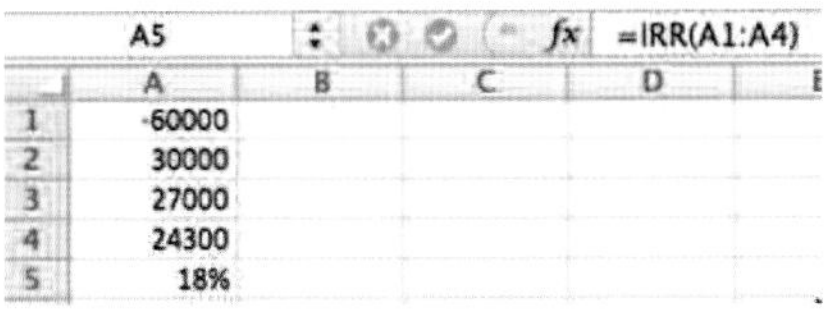

	A	B	C	D	E
1	-60000				
2	30000				
3	27000				
4	24300				
5	18%				

A5 ‖ ⊘ ⊙ *fx* =IRR(A1:A4)

Figure 10.3 Using Microsoft Excel to calculate the internal rate of return of an investment. The formula in the field at the top is entered into cell A5 and yields the percentage rate based on the figures above.

Source: Author

Ways of offsetting risk

Management may argue that the initial capital outlay for the project cannot be justified, despite the long-term benefits. In this case there are other possibilities. First, it is worthwhile considering asset finance such as leasing and renting. These techniques offset the monthly cost of the new equipment against the energy savings it delivers across the financing term, effectively making it a zero net cost or even cash-positive investment.

Second, it may be possible to obtain the services of an energy services company (ESCo), which would bear all the risk. This company then sells the service to the client under a long-term contract. The service could be heat, lighting, power, or a whole package which includes efficiency and energy management. In the USA, these companies offer energy savings performance contracts, under which they develop, install and arrange financing for improvements to boost energy efficiency and lower costs.

Third, land or roof space could be leased to a local utility or energy firm. They install equipment, for example, solar panels or a CHP plant, and sell the electricity, reaping the benefit of any tax credits or subsidies. They then pay rent to the client, and possibly sell a proportion of the energy generated at a discount. In the UK these arrangements come under the heading of power purchase agreements.

Finally, grants, tax credits and subsidies may be available at a local, national or federal level. In the USA, the Department of Energy's Energy Efficiency and Renewable Energy programme coordinates funding for projects. There are several tax credits, grants and rebates available. For example, a tax credit scheme is on offer until 2016 which gives 30 per cent credit on biomass, heat pumps, solar, wind and fuel cells. Energy-efficient mortgages are also available from the government and some private loan companies, which help buyers to renovate an inefficient property or build a new one. Different states offer different incentives, which vary widely. For instance, New York state offers grants of up to 50 per cent for renewable micro-generation. A good source of information is the Database of State Incentives for Renewables and Efficiency (DSIRE, at www.dsireusa.org). In some states there are electricity buy-back credits available for renewable energy fed back into the grid. Information is available from the local utility.

In the UK, the Low Carbon Buildings Programme offers grants for the public sector and non-profit organisations. The Carbon Trust is a good source of information on tax credits and other incentives. Feed-in-tariffs, Renewable Obligation Certificates and the Renewable Heat Incentive are other schemes. In 2018 these are due to be replaced by the carbon price floor scheme. Some projects may be eligible for finance from the Green Investment Bank at favourable rates, while specialist banks such as Triodos and the Cooperative Bank look especially favourably on schemes with environmental aspects.

Measurement and verification (M&V)

Measurement and verification (M&V) is the term given to the method of quantification of savings delivered by an energy efficiency measure. Measurement and verification demonstrates how much energy the measure has avoided using, a prerequisite for evaluating the total cost saved. There are four possible ways of doing this:

1   With respect to actual measurements of a particular performance indicator, such as the power used by a particular appliance, set of appliances, lighting, etc. that are being replaced, plus estimates, based, say, on models or historical data, and all other variables.
2   Based on actual measurement of all performance indicators.
3   Based on the actual measurement of energy use of a whole facility, or sub-facility that is affected by the measure. This is likely to require linear regression analysis with respect to outdoor air temperature, as described in Chapter 7.
4   Based on modelling or simulation of the energy use of the facility or sub-facility. This is employed where there is no historical data.

Evaluations must be as accurate as possible; for example, there may be additional electrical costs associated with the use of heating equipment. Transport fuel costs may need to be taken into account. Care should be taken to make sure that reporting is not over-optimistic. Transparency should be the watchword.

The International Performance Measurement and Verification Protocol (IPMVP), managed by the Efficiency Valuation Organisation, an international non-profit organisation dedicated to supporting M&V, is the globally accepted standard for quantifying the results of energy efficiency investments, demand management and renewable energy projects. It was created as an initiative of the United States Department of Energy. IPMVP is used to measure actual results against those modelled, and benchmarks.

ISO 50001, the internationally accepted energy management system standard, requires that an M&V plan be put in place and followed for any measure. In addition, the Efficiency Valuation Organisation runs accredited training in many countries in association with the Association of Energy Engineers leading to the qualification Certified Measurement and Verification Practitioner Professional (CMVP). This means that third-party independent evaluation of a project can be made if required, for example, for reporting purposes (CSR and annual reports).

Whether there is independent evaluation or not, the presence of a plan helps to preserve confidence in the investment that is to be made. If an energy manager wishes to remain successful and to build their reputation within the organisation, and, indeed of the organisation, they must build up trust, and this is one way to do so. The promised savings must actually be either delivered or exceeded. That is the bottom line.

It follows that the cost of M&V needs to be built into the cost of the project. If, for example, a BEMS is already installed, it forms part of the overheads. If the project includes the installation of a BEMS, then part of the payback will be the benefit of being able to monitor further projects.

Carbon strategy

Most companies now have a carbon-saving strategy with targets for reduction alongside financial targets. They need to record and measure their carbon emissions, and predict outcomes. Most of the BEMS and software described in Chapter 1 will have the facility to convert energy usage into carbon emissions, although data will have to be brought in from elsewhere as well, for example, transport. The energy manager's case to management must therefore include quantified energy savings and projections, just as it does financial savings. The carbon emissions abated by various measures vary according to the mix of supply in the local grid at that particular location.

Some organisations present CO_2 emissions in tonnes of carbon instead of tonnes of CO_2. To convert from tonnes of carbon, multiply by 44/12, which is the molecular weight ratio of CO_2 to C.

Political strategy

Like any senior manager, an energy manager needs to be a consummate strategist. He or she is competing for funds for projects with other managers in other departments. There are several tactics that may be used:

- Cultivating an ally on the board.
- Building up trust by first proposing no-cost or low-cost projects and quick wins.
- Not proposing any project which has not been thoroughly investigated and costed, and discussed with the people who would be affected by it.
- Looking at which departments or sites have the most potential for energy savings.
- Having several projects available at any one time to propose, with different advantages.
- Being able to demonstrate or conduct visits to successful similar projects elsewhere.
- Being able to demonstrate a competitive advantage to be gained from the project. This could be qualitative as well as quantitative, such as improvements in working conditions or an improved reputation for the organisation.
- Allying the project with the aims of the organisation as a whole, for example, the attainment of corporate targets. This might include setting an overall publicly declared energy target for the organisation.

The value of site visits

Financial reasons are not the only potential barrier to take-up of low carbon technologies. One of the best ways found in the extensive research on this subject to persuade people to take action is if they see a particular measure successfully employed elsewhere. Energy managers may find it beneficial to organise trips to such sites as exemplars of measures they have in mind, so that the reality may be brought home most dramatically. The testimony and recommendation of those who have already taken these steps, and the ability of visitors to question them to allay any doubts and dispel any misconceptions, are invaluable.

Note

1 Energy Use in Buildings and Industry: Technical Appendix, Mark Weiner, Committee on Climate Change, London, March 2009.

11

Conclusion

This book has suggested a host of ideas for reducing energy use in buildings. It is perfectly possible for a zero or even negative carbon building to be built or renovated using the approaches suggested. (A negative carbon building is one that sequesters carbon rather than emitting it.) We have also tried to give an indication of tactics that energy managers can use to evaluate them and how to mount a campaign within the business, both among staff and for the benefit of senior management, to get everyone on side. We recommend obtaining the professional qualification associated with energy management, ISO 50001, or any of the other standards.

It should be clear by now that energy management is not simply a matter of installing a building energy management system (BEMS), or an instrument control and automation system (ICA), or even a supervisory control and data acquisition system (SCADA). Energy management is a set of interrelated procedures, processes and practices. It requires a clear policy on the use of energy throughout the organisation whose targets are to reduce costs, overall carbon emissions and energy use to specific levels. It is essentially about inspiring people and assigning responsibility, and about a continuous monitoring and review process to develop the plan in tandem with changing conditions.

But it may still not always be clear exactly what measures should be implemented first. How can it be determined which ones will bring the most benefits? Only by crunching the numbers. The answers are not always in line with expectations, or what is conventionally thought of as green thinking. It may seem heretical, but it is not always a good idea to put in a wind turbine, although it might make the organisation appear to be 'green' (and there may be other benefits). This conclusion emerges from examining Table 11.1,[1] presented as an example, which compares various options for 'greening' existing inefficient lighting and considering which will be the most cost-effective:

Table 11.1 Comparison of costs and carbon savings from different strategies on lighting in a hypothetical case (thanks to Kit Oung)

Strategy	Install occupancy sensors to turn lights off when not in use	Replace existing lighting with T5 twin lighting with integrated PIR	Replace existing lighting with LEDs	Replace existing lighting with LEDs powered by wind turbine
Electricity saved (kWh/yr)	40,768	49,562	53,226	53,226
CO_2 abated (tonnes/yr)	18	22	24	32
Capital cost (£)	10,000	20,000	23,0001	162,000
Simple payback (years)	2.65	4.37	4.67	25.02
Cost of abated CO_2 (£/tonne)	545	897	970	5137

Cost-effectiveness may not be the main metric, but it's quite clear that just by turning off the lights when they're not necessary gives the largest saving: 1.8 tonnes of carbon dioxide. It would be even cheaper if the lights were turned off and on manually. The remaining savings are incremental, but the cost substantially increases once the wind turbine is put in.

Lifetime carbon emissions

Let us look at this from another angle, namely the lifetime greenhouse gas emissions savings of a given energy efficiency measure, with the help of a visual method of presenting these savings in order to compare them. Let us start with the example of a basic gas boiler (Figure 11.1).

Time is proceeding from left to right along the X-axis. Above the X-axis are carbon savings and below the X-axis are carbon emissions. Where the X-axis crosses the Y-axis is the beginning of the implementation of the measure. Any emissions made or saved before this point arrive as 'embodied carbon' with the product or service. The boiler begins work already responsible for some carbon emissions produced during its manufacture, delivery and installation. It then emits more climate warming gases during its lifetime of operation, and then causes more at the end of its life, when it is dismantled, recycled or whatever. The sum of all of these emissions, in the brown area, represents its lifetime carbon emissions, most of which occurred during its useful life, which is generally around 15 years. If we had reliable figures that would allow us to quantify all of these data for a particular case, we could determine the total by calculating this area mathematically.

Now compare this to a CHP boiler, i.e. the same boiler to which has been added heat recovery (see Figure 11.2).

Carbon emissions over lifetime of measure

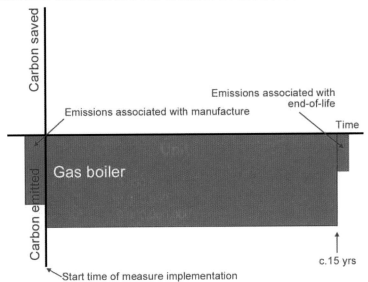

Figure 11.1 Imagined representation of the lifetime carbon emissions for a gas boiler.

Source: Author

Carbon emissions over lifetime of measure

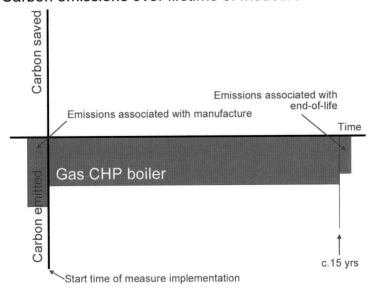

Figure 11.2 Imagined representation of the lifetime carbon emissions for a gas CHP boiler.

Source: Author

Here, the emissions associated with the beginning and end of the boiler's life are about the same, but emissions have been saved by capturing its otherwise wasted heat, because the heat did not have to be sourced elsewhere. Therefore the brown area below the line is not as deep.

Suppose the gas came from a renewable source, like anaerobic digestion. Then, all of the emissions associated with its operation would disappear. We could either credit these saved emissions to the boiler, and move the brown rectangle below the line to above the line, or we could credit them to the anaerobic digester. The graph for the latter would look like that in Figure 11. 3.

Here, there are substantial savings above the line, due both to the emissions saved by generating a renewable gas, and emissions saved by preventing the natural rotting of the organic waste, by normal composting or from landfill, which would produce methane, a much more powerful greenhouse gas than carbon dioxide. Subtracting the emissions caused at the end and beginning of its life would give us the total lifetime carbon emissions.

Figure 11.3 is similar to the next one, Figure 11.4, which represents emissions associated with installing any type of equipment, such as variable-speed drives or voltage optimisers, or adding fossil-fuel-based insulation to a building. In each case, there are below-the-line costs associated with manufacture and end of use, and above-the-line savings which will be lesser or greater according to the measure taken.

This is to be contrasted with the use of insulation materials derived from plant sources, such as wood fibre board, hemp, wool, cork and flax (see Figure 11.5). These have already absorbed atmospheric carbon that is now sequestered within them. Installing them in a building ensures that the 'embodied carbon' will stay out of the atmosphere for the life of the building or installation. Thus there is extra credit above the line, and the measure has hardly any carbon cost. At the end of its life, the insulation materials can even be fed into an anaerobic digestion system or composted.

It is interesting to think about Figure11.4 as representing the case of the installation of renewable electricity-generating plant such as photovoltaic modules or wind turbines. These also have carbon costs associated with their manufacture and disposal at the end of life. There is a term describing the point at which such

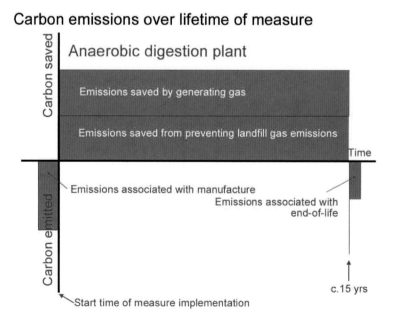

Figure 11.3 Imagined representation of the lifetime carbon emissions for an anaerobic digestion plant.

Source: Author

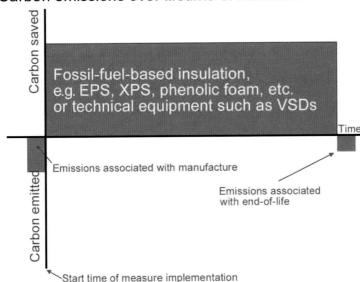

Figure 11.4 Imagined representation of the lifetime carbon emissions for fossil-fuel-based insulation or energy-saving equipment.

Source: Author

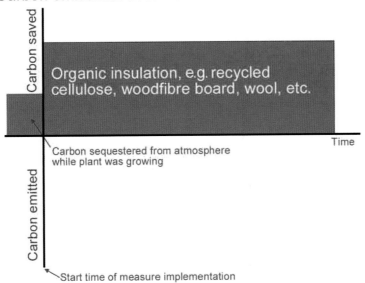

Figure 11.5 Imagined representation of the lifetime carbon emissions for organic insulation.

Source: Author

a piece of kit has generated sufficient renewable electricity to compensate for the carbon emissions created by its manufacture, which is its 'carbon payback period'. For a wind turbine this may be around nine months (similar to a nuclear power station) and for a PV module it may be between 18 months and three years depending on a number of factors. We can demonstrate this in Figure 11.6, moving the emissions behind the Y-axis to the right, and letting it cross the line at the point where the carbon payback period is reached.

Figure 11.6 Imagined representation of the carbon emissions payback for renewable electricity generators.

Source: Author

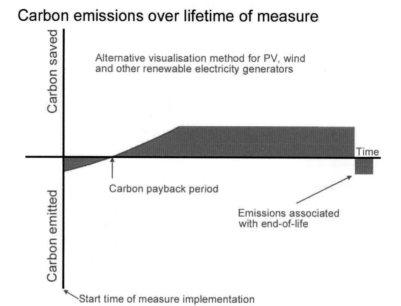

Carbon emissions over lifetime of measure

Alternative visualisation method for PV, wind and other renewable electricity generators

Carbon saved

Carbon emitted

Time

Carbon payback period

Emissions associated with end-of-life

Start time of measure implementation

Biomass boilers are an exception to this (see Figure 11.7). They are claimed to be carbon neutral because biomass such as trees or coppice on short rotation are replanted or regrown to replace that which is burnt. However, this only occurs after the fuel is burned and emissions have entered the atmosphere. If the fuel is short rotation coppice or pollarding, it could be, say, four years before the emissions are recouped, during which time the emissions are increasing global warming. If trees are allowed to grow completely before felling it could take 20 years. Of course, this process (burning and replanting) is happening continuously, and so the representation below is a 'snapshot' of a single burning. We will also see that there is a net negative balance, due to emissions associated with the manufacture and end of life of the boiler.

So far, we have only found one measure with no below-the-line carbon costs. But there are plenty more, and it turns out that they are the simplest, and the same as those we found above with the first exercise. They are low-cost or behaviour-change measures: changes which require the installation of no-cost, or inexpensive, pieces of equipment (see Figure 11.8). As well as obvious things like turning off machinery when it is not used and blocking holes through which draughts may enter, this also includes careful maintenance and supervision to make sure that equipment runs at its optimum efficiency and is not wasting energy or having its life shortened.

Eco-minimalism

There is another, more subtle conclusion to be drawn from this: that technology is only part of the solution. The rest is down to people and organisational culture. Engaging the workforce and management in the campaign to reduce energy consumption has very little capital cost but achieves much benefit. The most

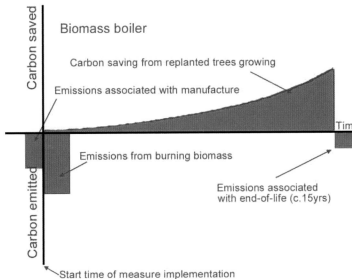

Carbon emissions over lifetime of measure

Biomass boiler

Carbon saving from replanted trees growing

Emissions associated with manufacture

Emissions from burning biomass

Emissions associated
with end-of-life (c.15yrs)

Start time of measure implementation

Figure 11.7 Imagined
representation of the lifetime
carbon emissions for a biomass
boiler.

Source: Author

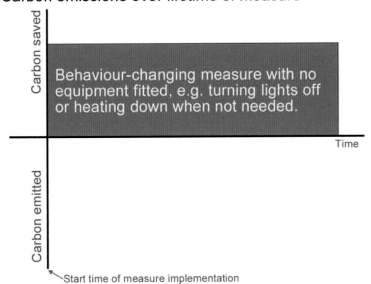

Carbon emissions over lifetime of measure

Behaviour-changing measure with no
equipment fitted, e.g. turning lights off
or heating down when not needed.

Start time of measure implementation

Figure 11.8 Imagined
representation of the lifetime
carbon emissions for a
behaviour-change measure with
no equipment fitted.

Source: Author

obvious energy reduction opportunities are often the most overlooked, and this
is down to personal responsibility. Energy consultant Kit Oung lists some of these
as follows:

- Lighting.
- Excessive ventilation compared to demand.
- Simultaneous heating and cooling.

- Idle time in production.
- Generation of graded or waste products.
- Poorly executed maintenance resulting in subsequent re-failure.

Oung notes that there are other benefits from implementing such measures. For example, turning down the air supply can make the workplace quieter. Improving lighting, in particular by adding daylighting, makes for a more pleasant working environment. Reducing the energy consumption of machinery can make it last longer. The energy manager will therefore look for multiple wins for this type of measure and include them in their advocacy presentations. Building efficiency expert Nick Grant calls this eco-minimalism.

He says that the question 'How do we achieve a zero carbon building?' should lead to questions like: 'Is it the right system boundary?' And even 'Do we need this building?' He says, 'adding extra insulation and renewable energy systems to compensate for an excessive carbon footprint is chasing our tail in environmental terms'. As Paul Anderson said, 'there is no problem, no matter how complex, which, if looked at in the right way, cannot be made even more complex'.

It is very important, therefore, for the right energy efficiency campaign to be conducted, and for good tools to be used in planning, such as the Passivhaus Planning Package (PHPP) for building design. This allows designers to optimise for minimal energy consumption and optimum comfort. It is equally important that actual performance of the results should be measured against predictions so that there is a possibility of learning from mistakes and correcting them. It has been suggested that energy use in non-domestic eco-buildings is typically around three times that predicted by the design. That is why we have been at pains in this book to recommend taking advantage of free energy and minimising its necessity in the first place.

Nick Grant is fond of quoting Emerson on Henry David Thoreau's attempt to live a self-sufficient life: 'He chose to be rich by making his wants few'.

Energy efficiency represents hidden wealth. It takes a great energy manager to reveal it. It is not just the organisation for whom they work, but the planet that will thank them.

Note

1   Energy Efficiency Consideration in Business: A Business Manifesto, K. Oung, Camfil, November 2012.

Appendix

Energy use

Units

kilo-	k	10^3	10,000
mega-	M	10^6	10,000,000
giga-	G	10^9	10,000,000,000
tera-	T	10^{12}	10,000,000,000,000
peta-	P	10^{15}	10,000,000,000,000,000

Example

milliwatt (mW): 1000th of a watt
kilowatt (kW): 1000W
megawatt (MW): 1,000,000W
gigawatt (GW): 1,000,000,000W
terawatt (TW): 1,000,000,000,000W. In 2006 about 16TW of power was used worldwide.

Abbreviations

Btu	=	British thermal unit (MBtu = millions of Btus)
MJ	=	Megajoule
TJ	=	Terajoule
Gwh	=	gigawatt-hours
toe	=	tonnes of equivalent oil (Mtoe = millions of toe)
Kcal	=	kilo calorie
Gcal	=	giga calorie

Conversion factors

To	TJ	Gcal	Mtoe	MBtu	GWh
From	**Multiply by**				
terajoule (TJ)	1	238.8	2.388×10^{-5}	947.8	0.2778
gigacalorie (Gcal)	4.1868×10^{-3}	1	10^{-7}	3.968	1.163×10^{-3}
million tonne of oil equivalent (Mtoe)	4.1868×10^{4}	10^{7}	1	3.968×10^{7}	11630
million British thermal unit (MBtu)	1.0551×10^{-3}	0.252	2.52×10^{-8}	1	2.931×10^{-4}
Gigawatt-hour (GWh)	3.6	860	8.6×10^{-5}	3412	1

From	To kWh. Multiply by
Therms	29.31
Btu	2.931×10-4
MJ	0.2778
Toe	$1.163 \times 10$4
Kcal	1.163×10-3

Example

Conversion of 100,000 Btu to kWh:

$$100,000 \text{ Btu} = 100,000 \times 2.931 \times 10^{-4} \text{ kWh} = 29.31 \text{kWh}$$

Conversion factors for mass

To:	kg	T	lt	st	lb
From	**multiply by**				
kilogram (kg)	1	0.001	9.84×10-4	1.102×10^{-3}	2.2046
tonne (t)	1000	1	0.984	1.1023	2204.6
long ton (lt)	1016	1.016	1	1.120	2240.0
short ton (st)	907.2	0.9072	0.893	1	2000.0
pound (lb)	0.454	4.54×10^{-4}	4.46×10^{-4}	5.0×10^{-4}	1

Conversion factors for volume

To	gal US	gal UK	bbl	ft3	l	m³
From:	multiply by					
US gallon (gal)	1	0.8327	0.02381	0.1337	3.785	0.0038
UK gallon (gal)	1.201	1	0.02859	0.1605	4.546	0.0045
barrel (bbl)	42.0	34.97	1	5.615	159.0	0.159
cubic foot (ft³)	7.48	6.229	0.1781	1	28.3	0.0283
litre (l)	0.2642	0.220	0.0063	0.0353	1	0.001
cubic metre (m³)	264.2	220.0	6.289	35.3147	1000.0	1

Carbon dioxide emission factors by gross calorific value

Energy source	KgCO$_2$/kWh	KgCO$_2$ per other units
Natural gas	0.18523	5.3808 per therm
LPG	0.21445	6.2915 per therm
Coal	0.32227	2383 per tonne
Diesel	0.25301	3188 per tonne
Petrol	0.24176	2.3117 per litre
Fuel oil	0.26592	3228 per tonne
Burning oil	0.24683	3165 per tonne
Wood pellets	0.03895	183.9 per tonne

Note: The greenhouse gas conversion factor comprises the effect of the CO2, CH4 and N2O combined, quoted as kgCO²e per unit of fuel consumed.

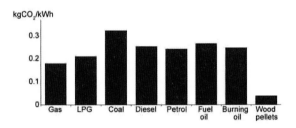

Figure 12.1 Comparison of the global warming potential of various fuels, showing kilograms of CO² produced per kWh of energy generated.

Source: Author

The global warming value for electricity varies by time, country and location, depending on the fuel mix. Figures are available from the Independent Energy Agency (http://www.iea.org/publications/freepublications/publication/name,4010, en.html).

Power and energy

Power is the rate at which energy is produced by a generator or consumed by an appliance.

Unit: the watt (W). 1000 watts is a kilowatt (kW).

Energy is the amount of power produced by a generator or consumed by an appliance or over a period of time

Unit: the watt-hour (Wh). 1000 watt-hours is a kilowatt-hour (kWh), commonly a unit of electricity on a bill.

Alternate unit: the joule (J). Watt-hours can be used to describe heat energy as well as electrical energy, but joules are also used for heat. 3600 joules = 1Wh. Put another way, a joule is 1 watt per second, since there are 3600 seconds in an hour; or 3.6 megajoules (MJ) = 1kWh.

Examples

- One 80W light bulb on for two hours, or two 80W bulbs on for one hour would consume 2 × 80 = 160Wh.
- Three 80W light bulbs on for six hours will consume 3 × 80 × 6 = 1440Wh or 1.44kWh.

Energy efficiency and lighting

For the same amount of luminescence over four hours:

- An 80W incandescent bulb will consume 80 × 4 = 320Wh.
- A low energy 18W compact fluorescent bulb will consume 18 × 4 = 72Wh.

From this we can see that a compact fluorescent bulb is 72/320 = 4.5 times more efficient than an incandescent bulb.

Fans and air conditioning

Fans consume less energy than air conditioners. A typical fan of 30W on for six hours will consume 30 × 6 = 180Wh. Conversely a typical air-conditioning unit of 2kW or 2000W on for the same period will consume 2000 × 6 = 12,000Wh. That is 67 times more energy.

Power generation

- One photovoltaic solar panel producing 80W for two hours, or two panels producing 80W for one hour would produce $2 \times 80 = 160$Wh.
- Three panels producing 90W for five hours will produce $3 \times 90 \times 5 = 1350$Wh or 1.35kWh.

Insulation

There is a relationship between the thermal conductivity of any material, its thermal resistance (the R-value) and its heat transfer (the U-value) properties, which all relate to the standard of insulation we want for a low carbon building.

Thermal conductivity (k)

Thermal conductivity, k (also known as psi or denoted λ), tells us how well a material conducts heat. The figures are supplied by manufacturers. It is:

$$k = Q/T \text{ times } 1/A \text{ times } x/T$$

or the quantity of heat, Q, transmitted over time t through a thickness x, in a direction perpendicular to a surface of area A, due to a temperature difference T. The units used are either SI: W/mK or in the US: Btu/(hr $\times$ ft $\times$ °F). To convert, use the formula 1.730735 Btu/hr $\times$ ft $\times$ °F = 1 W/mK.

R-value

The R-value is a measure of how well a material resists heat travelling through it. It is the ratio of the temperature difference across an insulator and the heat flow per unit area through it. The bigger the number the better the insulator. It is the depth/thickness of a material divided by its thermal conductance; in other words, R = d/k.

To compare two insulants with different thickness and thermal conductivity, it is necessary to calculate the value of R for each.

R-values are given in metric units: square-metre Kelvin per watt or $m^2 \times K/W$ (or equivalently to $m^2 \times °C/W$); or, in the United States, in $ft^2 \times °F \times h/Btu$. It is easy to confuse them because R-values are frequently cited without units (e.g. R-3.5). One R-value (US) is equivalent to 0.1761 R-value (metric), or one R-value (metric) is equivalent to 5.67446 R-value (US). Usually, the appropriate units may be inferred from the context and their magnitudes.

Doubling the thickness of an insulating layer doubles its thermal resistance. R-values are often used when there are multiple materials through which heat can travel. The R-values of adjacent materials may be added together to calculate the overall value (e.g. R-value (brick) + R-value (insulation) + R-value (plasterboard) = R value (total)). Another way of calculating it is to add the inverse of the k values of each element multiplied by their thickness, or: R(total) = (1/k) × d. Remember to include internal and external resistances, and unvented air gaps.

U-value

R-value is the reciprocal of U-value (and vice versa of course). A lower U-value indicates greater insulation value. Commonly used in Europe, it is the overall heat transfer coefficient, describing the rate of heat transfer through a building element over a given area, under standardised conditions. The usual standard is at a temperature gradient of 24°C, at 50 per cent humidity with no wind.

It is described in watts per square metre Kelvin ($W/(m^2 K)$) or the amount of energy lost in watts per square metre of material for a given temperature difference of 1°C or 1°K from one side of the material to the other. Another way of understanding it is to see it as thermal conductivity divided by the depth of insulation, or U = k/d where k is the thermal conductivity of a material, d the materials depth.

Building regulations provide minimum standards of thermal insulation, typically expressed as a U-value for a given building element. This is found by adding the U-values for the different materials times the depth and area used for each within the element. In each case, measurements are taken on-site and then reference is made to information tables for the purpose of the calculation.

Calculating the R-value or U-value of an entire building

To calculate the R-value of a complete wall, it's necessary to add the U-values of each section (e.g. parts containing studs and those parts without studs, lintels, etc.) multiplied by the percentage of the overall wall they represent, and then inverse the result. This process is repeated for each building element (walls, roof, floors). These complex calculations are undertaken with bespoke software such as the free one at www.thermalcalconline.com.

Calculating heat losses and gains

There are two primary methods of heat loss or gain in building: conduction through the building envelope, and air movement through and between the elements.

The general heat loss/gain formula is: Q = U*A*ΔT, where the heat loss of an area of size A is determined by the U-value of the materials and the difference in temperature between inside and out (that is the difference in temperature of the two surfaces, not the two air temperatures).

Index